The World of Dinosaurs

Exeter Books
NEW YORK

First published in USA 1983
by Exeter Books
Distributed by Bookthrift
Exeter is a trademark of Simon & Schuster, Inc.
Bookthrift is a registered trademark of Simon & Schuster, Inc.
New York, New York

ISBN 0 671 06461 4

Printed in Czechoslovakia

The World of Dinosaurs

Written by John Heritage

Illustrated by W. Francis Phillipps
and other artists

CONTENTS

The never-ending story

Through the swirling dawn mists of a steaming swamp lurched several long-necked monsters. Not one was less than 21 metres (70 feet) long, not one weighed less than the combined weight of five present-day elephants. They would seem awesome and terrifying to us if they were alive today, but this was a dawn some 230 million years ago, long long before any human being walked the Earth.

Though they were gigantic creatures, they were not particularly dangerous to the other inhabitants of their world for these huge beasts did not kill other animals for food. They were plant-eaters and today they are probably the best-known of all the dinosaurs.

Since the bones of the first specimen were discovered, this long-dead animal has been called Apatosaurus, which means 'unreal lizard'. Possibly the Apatosaurus was so named because the man who found its skeleton could not believe it was real. This animal is more popularly known as the Brontosaurus, which means 'thunder lizard' for it is reasoned that 'When that monster strode the Earth, the noise of its passing must have sounded like thunder.'

If the Apatosaurus was not flesh-eating, the Allosaurus certainly was. This dinosaur was the biggest flesh-eater of its time and its main source of food was the flesh of the long-necked Apatosaurus. The Allosaurus (its name, oddly, means 'other lizard') was only half the size of an Apatosaurus but this did not prevent the savage beast from launching a merciless attack on its intended prey.

As these gigantic long-necked dinosaurs plod along, munching the tasty plants growing profusely in the swampland, an Allosaurus suddenly pounces on a young Apatosaurus. The victim collapses, already writhing in its last agonies, for its enemy inflicts an immediate death wound before the Apatosaurus can make a single move to defend itself. This incident is pictured on pages 8 and 9. The Allosaurus does not escape completely unscathed for so savagely and violently does it bite into the still heaving body of its prey that some of its teeth are torn from its jaws.

If you think that this incident is entirely a figment of imagination, then you are wrong, for when the skeleton of that Apatosaurus was found not so long ago, the broken teeth of the Allosaurus were found among the bones. The discovery graphically informed the scientists that the Allosaurus was partial to a little Apatosaurus meat.

It is, of course, from such finds that learned men have been able to piece together so much information about the strange, mysterious creatures who lived when our world was millions and millions of years younger.

On this page is a photograph of the reassembled skeleton of another dinosaur known as Triceratops, which translated means 'three horned face'. You will be able to read more about this plant-eater later in the book.

New finds are constantly being made by palaeontologists (the name given to scientists who study extinct forms of life). A new find may confirm an already established

This giant skeleton of a Triceratops gives some idea of how awesome the beast was. This monstrous animal was the biggest of the Ceratopsian dinosaurs and lived during the period known as Cretaceous. Despite its terrifying arrangement of horns, one above each eye and one on its nose, which must have made it awesome to the other animals then living it was a harmless plant-eater and did not kill in order to eat the flesh of its victims.

This is the skeleton of an Ichthyosaur, a fish-reptile which had a tail similar to that of a modern shark. The fin on its back was similar in shape to that of a present-day dolphin's. Most marine reptiles come ashore to lay their eggs. Not the Ichthyosaurs! Their eggs hatched inside the mothers' bodies and fossils of baby Ichthyosaurs inside their mothers have been found. Baby Icthyosaurs were born live into the sea, but despite its appearance, the Ichthyosaur was not *a fish. It was a reptile, whose legs became powerful fins over a period of millions of years. These enabled it to swim long distances. A hard ring of flesh surrounding its eyes enabled Icthyosaurs to dive to great depths as the ring prevented great pressure of water damaging the eyes.*

theory concerning some long-dead prehistoric animal or, occasionally, a discovery may destroy a much-cherished notion. For example, it has recently been reported in *The Times* newspaper of London that the first discovery in South America of dinosaur bones from the Jurassic period of 150 million years ago has been made in Patagonia, Southern Argentina, and that they promised to shed new light on a period of evolution about which relatively little is known. It would seem that this find supports the theory that certain dinosaurs migrated between North and South America during the Jurassic period. Also, although many of the bones resemble those found in North America, some are much more primitive. Comparison with the later skeletons will help paleontologists trace the evolution of the dinosaurs.

The Times also reveals that two palaeontologists in the USA have assembled overwhelming evidence that the head of the Apatosaurus is different from that previously held to be true. It seems that instead of having a blunt face and jaws containing flat spoonlike teeth, the head was in fact long and pointed, with pencil-like teeth. Apparently, for the last hundred years, palaeontologists have believed that something was wrong with the traditional idea of the shape of the head but lacked the evidence that is now forthcoming. Museums all over the world are planning to make the necessary changes to their models and drawings. The search for new facts continues. It is a story that never ends.

Skeletons and fossils, such as those illustrated on these pages, were the clues to the discovery of the dinosaurs. A bone dug here and a footprint found there added pieces to the jigsaw of discovery. When it was revealed in the 19th century that the Earth was once populated by monsters, scientists and palaeontologists set out in search of the dinosaurs.

How astonished Dean William Buckland was when he dug into a slate quarry at Stonefield and found what he described as the remains of a giant lizard.

A massive jaw, part of a pelvis, a section of a shoulder blade and several large back bones were all part of a giant skeleton which must have once belonged to some prehistoric monster. Dean Buckland worked out that the creature must have measured about 15 metres (50 feet) long and approximately three metres (ten feet) high, classified it as a reptile and gave it the name of Megalosaurus (meaning 'large lizard').

Remains of such creatures had been found before this discovery, which was made at the beginning of the 19th century. The first hints of these long-buried monsters of the past had been found in various parts of the USA but there had never been

This is a skeleton of the first bird, known as Archaeopteryx. A descendant of the flying reptiles, the Archaeopteryx was the first flying creature to have developed the feathers necessary to the body's insulation as they prevented it losing heat or becoming too hot while flying at high or low altitudes. Because the Archaeopteryx had feathers, we deduce that during its development it had become warm-blooded and therefore no longer belonged to the species of cold-blooded reptiles.

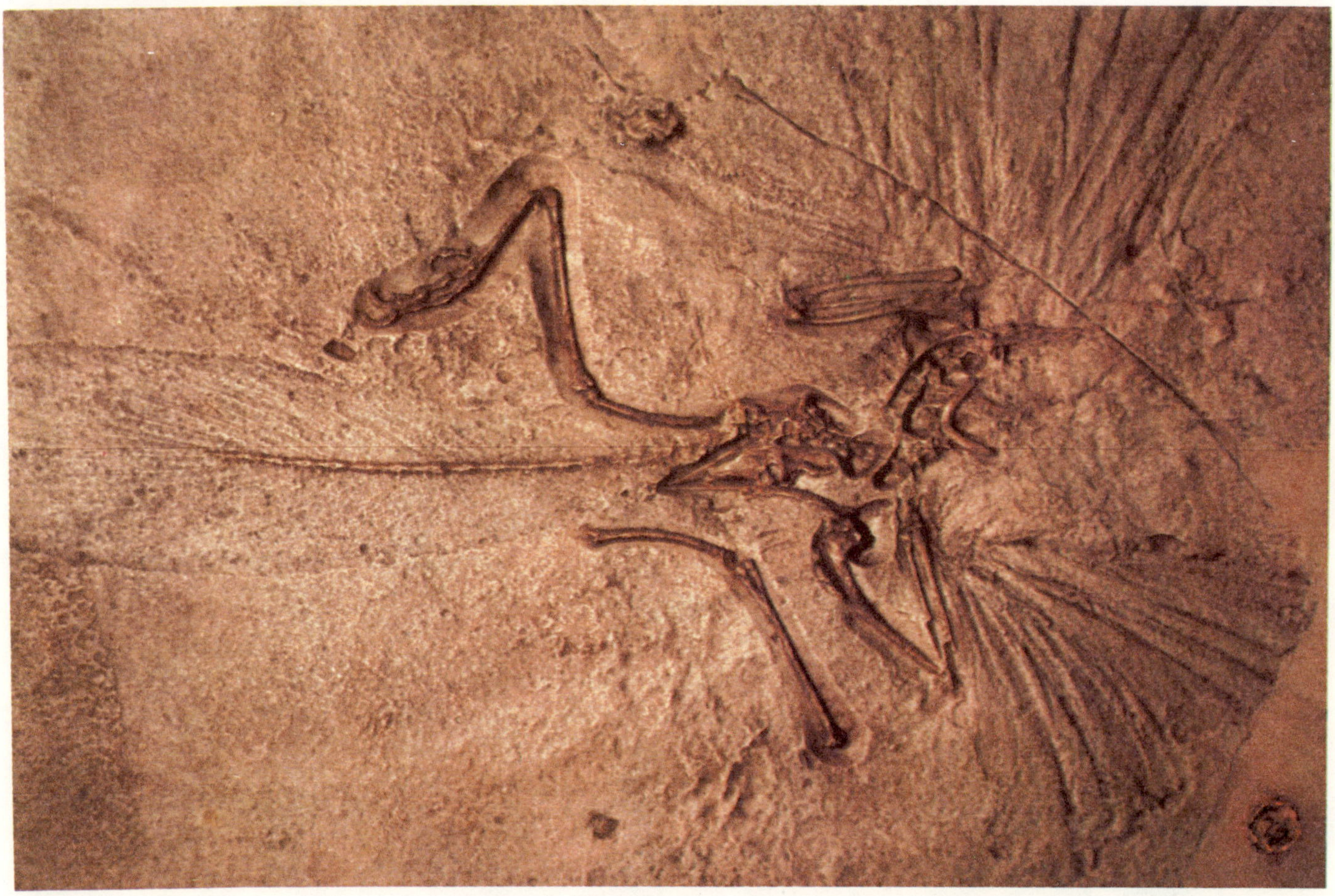

The chart shows the various periods in prehistoric times from the earliest forms of life to the threshold of man's appearance on Earth. It is truly a most fascinating and, at times, incredible record.

enough evidence to establish the fact that dinosaurs had ever really existed. No records were kept of these first discoveries, such as the thigh bone found in New Jersey in the 18th century, or the giant rib discovered on the south bank of Yellowstone River in 1806. These early finds were virtually ignored. Prehistoric monsters were found only in fairy tales, they surely had never existed.

But they had. The work of two English pioneers in dinosaur discovery did much to change this attitude. William Buckland and Dr Gideon Mantell became the founders of modern palaeontology, and to provide irrefutable evidence of the existence of such

GEOLOGICAL ERAS AND PERIODS

ERA	PERIOD	Millions years a
CENOZOIC	Quaternary	rece to 1.
CENOZOIC	Tertiary	63
MESOZOIC	Cretaceous Jurassic Triassic	135 180 225
PALAEOZOIC	Permian Carboniferous	270 350
PALAEOZOIC	Devonian Silurian Ordovician Cambrian	400 435 500 570
	Pre Cambrian	4500

monsters. Today palaeontologists have divided hundreds of millions of years into named eras and periods (see the chart opposite). We can state with authority for instance that the Diplocaulus (see above) lived in the Permian period.

In March 1822, Mantell and his wife were visiting a patient in Lewes, Sussex. It was here that the second dinosaur remains to be found in England were discovered by accident in a pile of rubble which had been put aside for road repairs. Mrs Mantell had

From 280 to 230 million years ago during the period known as Permian, Diplocaulus, an aquatic reptile, lived. A most extraordinary looking animal, it had a wedge-shaped head with weak legs. It was about 60 cm (2 feet) long.

noticed something glinting in the rubble heap and when she and her husband went closer to investigate, they found the glint came from some fossilized teeth embedded in a piece of stone.

Mr Mantell's curiosity was so aroused that he returned to the site for some weeks afterwards and to his great delight found a number of large fossilized bones, none of which he could identify, as well as more teeth. He decided to send them to Baron Cuvier in Paris, who was a famous authority on backboned animals. The Baron declared that the bones were those of an extinct hippopotamus and that the teeth belonged to an extinct rhinoceros.

Dr Mantell could not accept such an opinion and so he took the teeth to the Museum of the Royal College of Surgeons in London, where he compared the teeth with anatomical specimens of other animals. He soon found that they resembled the teeth of an Iguana, a lizard found in Central America, though the teeth he had

The big animal standing on its hind legs is an Iguanodon, so-named because Doctor Mantell who first found some of its fossilized teeth detected a resemblance between them and the teeth of the Iguana, an animal found in South America today. Mantell's discovery of dinosaur fossils was only the second in England and caused great excitement.

found were much larger. Mantell decided to call the creature whose teeth he had found an Iguanodon, which literally means 'Iguana-tooth'. On the opposite page is an illustration of an Iguanodon, together with a Phobosuchus, or 'horror crocodile', and a Seymouria.

The word 'dinosaur' was coined by Sir Richard Owen, a famous 19th century anatomist, as a general term to describe these prehistoric reptiles. Dinosaur means 'terrible lizard' and many of these creatures certainly lived up to their name!

By 1840 the remains of other creatures had been discovered. Some were found in England, others on the European continent. But until this time no complete skeleton of such a creature had come to light.

Then, in 1878, a great discovery was made in Belgium which proved beyond any shadow of a doubt that these monsters had existed. The discovery was made at Bernissart, in a coal-mine about 300 metres (1000 feet) below the ground when some miners stumbled upon a mass of fossilized bones when they were developing a gallery in one of the mines. Experts were immediately called in and for three years Monsieur de Pauw, the chief of the laboratory at the Royal Natural History Museum of Belgium, lived as a miner digging out bones and other Iguanodon remains until he had discovered more than 20 complete skeletons.

Millions of years ago these Iguanodons must have fallen into a ravine, and the problem of getting them out intact was a major one for the excavators. But they were carefully extracted and numbered, cleaned and rearranged, and finally exhibited to the public at the Natural History Museum in Brussels.

By examining these remains the experts were able to gain an extensive knowledge of what the Iguanodon looked like and in what conditions it lived.

So, by the 1880s the remains of two types of dinosaur had been found, and proof of the existence of such creatures had at last been firmly established.

There followed what can only be described as a great dinosaur 'rush' in America, and no one could have known how many exciting fossil finds lay in store for the American pioneers of palaeontology.

THE RIVAL PALAEONTOLOGISTS

For a quarter of a century, between 1870 and 1895, two of the most famous of these pioneers, Charles Marsh and Edward Cope, opened up the way to the immense fossil fields of the American West. Both men were ardent palaeontologists, both had inherited vast fortunes, and both were to transform palaeontology into a dynamic science.

The expeditions which they carried out were beset with difficulties from the start. Enormous distances had to be covered over trackless terrain, and the excavated gigantic heavy fossils had to be transported with great care to the nearest railway stations in unsteady wagons.

Marsh and Cope could have combined forces to form a dinosaur-digging team; instead they became bitter rivals, vying with each other until their respective expeditions became nothing less than a competition to see who would discover the most exciting and the greatest number of remains.

Unlike the earliest discoveries made in the United States during the 18th century in which a single bone was dug up in one area, and a different one dug up in another, the finds of Marsh and Cope were on a grand scale. In one area the remains were so plentiful that a cabin was built out of them. This was called Bone Cabin and eventually became the headquarters of the dinosaur 'dig' in that area.

To the south of Bone Cabin was Como Bluff, a long ridge in South Wyoming which runs parallel with the Union Pacific Railway. In 1887 two railroad men stumbled on a great mass of fossilized bones which they passed on to Marsh. Como Bluff soon revealed its long-hidden secrets. Complete skeletons of Jurassic dinosaurs were unearthed.

While Marsh had been working on this mass of remains at Como Bluff, Cope had not been idle. A bagful of enormous bones discovered in the Rockies at a place called Garden Park, Colorado, had been dispatched by a local collector to Cope, and so began the conflict and rivalry between the two palaeontologists.

A year earlier, in 1886, Cope had made a trek to Helena, Montana, where some Cretaceous remains had been discovered. The decision to make the expedition had been a difficult one for Cope, as the remains were in Sioux territory and the famous Indian Chief, Sitting Bull, had recently won the famous battle of Little Big Horn in which General Custer and his forces had been destroyed. The Sioux were reputed still to be in a very warlike mood. Luckily, however, the Sioux gave Cope no trouble and the trek proved to be a very profitable one.

During this expedition through Sioux country a new method of preserving dinosaur remains was first employed by Cope. Strips of cloth were soaked in rice boiled to a thick paste and wrapped around the bones. Later the method used to set broken limbs was adopted. This involved using porous cloth soaked in plaster of Paris and is still in use today.

In 1892, a few years before his death, Cope returned to Sioux country to look for Cretaceous remains and found, among other fossils, bones of the Duck-billed dinosaur, or Hadrosaurus, which means simply, 'big dinosaur'. The body of this creature was usually about 10 metres (15½ feet) long and weighed about three tonnes. The shape of its head and jaws has suggested the name 'duck-billed'. The jaws were broad and flat in front and were probably covered by a duck-like horny bill. Cope was richly rewarded for making such a difficult journey for the second time, by finding a magnificent skeleton of one of these dinosaurs.

It certainly looked as though Cope had beaten his rival. He had found a wonderfully rich hunting ground. But he did not live long enough to continue his successful work.

He had squandered his great fortune by foolish speculation in a vain attempt to prove that he could do better than the more wealthy Marsh. He died only a few years after his second expedition, leaving his rival to attain fresh honours in the field of dinosaur discovery.

Charles Marsh also turned his attention to Cretaceous remains towards the end of his career. He discovered the armoured dinosaur Nodosaurus and the more slender and graceful ostrich-like Ornithomimus.

The deaths of Marsh and Cope did not mark the end of the great American dinosaur rush. The North American continent was a particularly rich store of dinosaur remains and there was still much to be discovered. What they had done was to pave the way for the palaeontologists who were to come after them.

ENTER BARNUM BROWN

In the summer of 1893, a young American student at the University of Kansas, named Barnum Brown, led a fossil hunt through Nebraska and South Dakota.

It was never stated whether this expedition was successful or not, but the ones which followed it certainly were. In the following summer Brown led another fossil hunt, this time through Wyoming, where he unearthed a skull of a Triceratops. The skull of a Triceratops can be seen on page 7. Remains of this creature had been found before by Charles Marsh but for a young, inexperienced student such a discovery was quite a feat.

After graduating in 1897, Brown became associated with the American Museum of Natural History and was immediately sent on an expedition to Wyoming to collect dinosaurs. This was the beginning of his long and successful career in dinosaur-digging, during which he was to become one of the greatest discoverers of dinosaurs in America.

He collected many sorts of fossils, in the United States and throughout the world, but digging for dinosaur remains was Brown's chief joy. It was often said by those who had assisted him on his expeditions that he could almost 'smell' them out. The imposing collection of Cretaceous remains in the American Natural History Museum, which he served until his death in 1963, is a wonderful monument to his life's work.

One of Brown's greatest and most exciting discoveries was in the Howe Quarry at the foot of the Big Horn Mountains near Yellowstone Park. In 1934 a rancher named Barker Howe had reported a find of huge bones on his ranch and Brown was sent to investigate. He found what could almost be described as a dinosaur quarry. Bones were breaking through the surface of the ground, having weathered their way through the soil.

The bones were piled in heaps, all jumbled together, and what was even more interesting was that they were in a

This is the skull of the mighty Tyrannosaurus Rex (see page 19).

horizontal position. Obviously some natural disaster had occurred here millions of years previously.

A similar find at Utah showed the bones in a vertical position. Here, the land surface, originally flat, had been heaved upright. It was believed that the find in Utah had been caused possibly by a flood or a river in spate. The dinosaurs, some very large, had been swept along by it until they were stranded on a bank. The experts had come to this conclusion because the tails of all the dinosaurs were laid in the same direction, and while the upper parts of the skeleton were reasonably intact, the under parts had broken away.

But at Howe Quarry, as it was later called, there was nothing like that. The site of this ancient disaster had remained in a horizontal position. The bones lay in a confused state of disarray. It looked as though the dinosaurs, trapped somehow, perhaps by a flood, had trampled on each other. This theory was made even more plausible by the fact that there was a dense heap of bones towards the centre, thinning out towards the edges of the pile of bones.

One suggestion is that the disaster occurred in a narrow ravine when a sudden spate of water poured along it. The dinosaurs had therefore been trapped where the ravine sharply narrowed.

Another theory is that the dinosaurs had become stuck in the mud during a drought, and in a desperate effort to escape from it, had begun to trample on each other. That would account for the disorganized and broken state of the bones, but why the density of them towards the centre? The questions are endless, and many remain unanswered even to this day. This is one reason why the study of dinosaurs has become such an exciting and fascinating subject.

The highlight of Barnum Brown's career came in 1910 when he travelled down Red Deer River in Alberta in a barge-like craft, called a scow. This was built specially for the event and was large enough for the crew to erect a tent on deck as well as having the necessary cargo capacity to transport heavy fossils.

Styracosaurus was another horned dinosaur, whose appearance would have made a hungry carnivore think twice before attacking. It could use its great horn to devastating effect. The Ceratopsian group to which the Styracosaurus belonged were the last group of dinosaurs to evolve. They first appeared about half way through the Cretaceous period and survived in various forms for 35 million years. Despite its fierce appearance, the Styracosaurus was a herbivore, that is a plant-eating animal.

The expedition was a great success. By the end of the summer the scow was loaded with dinosaur remains and another trip was planned for the following year, working down-stream.

By the third year, Brown and his men had arrived at an older rock formation. Thus, specimens from different levels, or strata, were obtained. This made it possible to judge how these reptiles had developed during the course of millions of years.

The greatest of Brown's achievements was the discovery of two complete skeletons of the giant of them all: Tyranosaurus Rex, the Tyrant King, at the Hell Creek Formation of Eastern Montana. They are two of the finest dinosaur skeletons ever found and were a fitting crown to Barnum Brown's successful dinosaur-digging career.

The great dinosaur 'rush' had begun to die down by the turn of the century but in 1913 one more outstanding discovery was made in the North American continent. That year one of the horned dinosaurs, Styracosaurus (seen on these pages) was found by L M Lambe in Alberta in Canada. This monster, which weighed three tons, had a huge nasal horn and neck frill armed with six spines.

The horns of the Styracosaurus provided an effective defensive weapon like those of the Triceratops and Monoclonius. Armed with these, the horned dinosaurs were able to hold their own against some of their most powerful and fierce enemies.

There were other discoveries in North America at this time. The Sternbergs, a father and son team, carried on the work of Cope and Marsh the real pioneers who had led the great search for dinosaurs in America.

THE DISCOVERIES IN MONGOLIA

The North American continent was not the only rich storehouse of dinosaur remains in the world. Africa yielded wonderful results. Other parts of the world south of the equator also gave up their long buried secrets, but most of these discoveries were only a repetition of those which had been made already in North America.

One exception is the expedition across the vast empty space of desert scrub and grassland of Mongolia and the Gobi Desert, organized and financed from the United States.

Curiously, this expedition to Mongolia was planned with another end in view. Dinosaur hunting was no part of the plan at all.

At the turn of the century a theory was put forward which suggested the possibility that Mongolia was the area where primitive Man and mammals had first evolved.

The idea caught on although there was precious little evidence to go on, however. Only one fossil rhinoceros had been found in Mongolia, hardly enough proof to establish that country as the birthplace of modern life. Finally Roy Chapman Andrews, a scientist on the staff of the United States Museum of Natural History, led the expedition to Mongolia.

Great distances had to be covered in a few months, before the severity of the winter could begin to hinder the work of the scientists.

Four such expeditions were carried out in all, between 1922 and 1930.

The expeditions were unsuccessful in establishing Mongolia as the birthplace of Man. Other discoveries were made, which were similar to those made in the USA and Canada, including a complete skeleton of the mighty Tyrannosaurus.

The earliest horned dinosaur, the Protoceratops, was found in an area of giant sandstone. The most exciting part of this discovery, made at a place called Shaberakh Usa in the Gobi desert, in 1923, was that several batches of Protoceratops eggs were found. Near these nests were eighty skeletons of this dinosaur at various stages of development from newly hatched young, to adults measuring up to two metres (seven feet).

More was to come. Inside the eggs were ossified embryos, and nearby were the remains of a small two-legged reptile which has been named Oviraptor, meaning 'egg-stealer'. The Oviraptor must have been 'caught in the act' of stealing the eggs. How the event became ossified is a mystery. Perhaps a sudden violent sandstorm arose and buried the actors of this prehistoric drama, leaving their remains to be discovered by modern man centuries later.

When the first fossil skull of the Protoceratops was discovered, the scientists realized that this animal was related to the Ceratopsian dinosaurs. Ceratopsians (meaning 'horned faces') had beaked mouths and bony shields protruding from their heads and horns. The Protoceratops, did not *have a horn but the paleontologists believed that it resembled the Ceratopsians.*

The entry of the Reptiles

The surface of the Earth has not always been as it is today. It has been constantly changing through the thousands of millions of years since our world was formed.

What once were continents sank and became the beds of oceans and what once were seas became dry land. These changes occurred time after time. And each change resulted in new layers of rock or strata.

These changes resulted in the Earth's crust becoming like a series of sandwiches one on top of the other, each with different kinds of fillings. Found in each layer are the fossil remains of the animal and plant life that existed when that particular layer was formed and was on the surface.

Geologists, the scientists who study the history and formation of rocks, can tell at which period of the Earth's history any particular layer of rock was formed.

In that way it is possible to establish from the age of the rock in which a fossil is found the period when the animal or plant lived.

Large areas of many of the rock layers were quite soft when they were on top of the Earth's crust. Therefore it was easy for footprints, skeletons and other traces of animals to be left in them. Later, the ground containing the impressions hardened into rock. Later still, the impressions would be covered with further layers of rock.

The remains of animal or plant life had to be covered fairly quickly by movements of the Earth to be preserved as a fossil. There were a lot of earthquakes and volcanoes to cause the movements millions of years ago.

Shale, limestone and clay are the most likely rocks to contain fossils. Chalk, which is partly made up of the shells of tiny marine creatures which died millions of years ago, also contains many fossils.

By far the best conditions for preserving fossils were under the sea. That is why most fossils are found in what we call sedimentary rocks. There rocks were once the beds of oceans, but continual changes and volcanoes have thrust them upwards to form land and mountain ranges.

This does not mean that all the fossils found in sedimentary rocks are those of fish. Many of them are the remains of land animals which were swept out to sea and drowned during floods.

A

B

Occasionally the actual skeletons of animals that died hundreds of millions of years ago are found as fossils. This kind of fossil was produced when the animals were trapped in bogs and quickly buried.

It is much more common to find fossils in the forms of petrified skeletons. That is, skeletons that have been turned into stone. This happens when water from the surface of the ground seeps through carrying mineral matter which causes a chemical change which turns the bone into stone.

Coal is made up of the remains of plants and trees and for that reason some of the most perfect fossils of the world's ancient vegetation are found deep in coal mines.

Fossils range in size from the bone impressions of the giant dinosaurs to the remains of tiny shell-fish less than a centimetre across.

Besides telling us about the plant and animal life that existed millions of years ago, fossils give us a great deal of information about the changes of climate that have occurred in the course of thousands of centuries.

Fossils of animals and plants which could have lived only in the tropics have been found in Antarctica. Therefore we can be certain that where there is now snow and ice, there was once tropical heat and jungle.

Then there are the fossils dug up in Europe, which prove that elephants and rhinos once roamed throughout Britain and the countries of Northern Europe. Other fossils prove that most of Europe and North America were long ago covered with steaming swamps and tropical forests.

Only a few of the plants and animals of the past left fossil remains. Many factors prevented fossils forming, and it has been estimated that 99 per cent of all living things of the past have perished without leaving any trace behind.

Today the study of fossils is an exact science and gives us a reliable picture-story of what the world and the life on it were like over periods of millions of years.

We know from their evidence that millions of years ago huge, lizard-like animals, weighing 80 or more tons and up to 30 metres (98 feet) long, roamed the earth. They prove that dragonflies with a wingspan of a metre (three feet) flitted along the banks of rivers that dried up thousands of centuries ago.

We know that there were huge 'birds' with teeth in their beaks and covered with scales instead of feathers. And that plants and trees were bigger than any seen today.

Although the last of the fantastic creatures called dinosaurs and the strange plants, birds and insects with which they shared the world, died hundreds of millions of years before Man was on the Earth to see them, we know that they existed from their fossils – the stone 'portraits' of themselves they unwittingly left behind them.

C

Three examples of prehistoric fossils.

A. *Dinosaur eggs.*
B. *A Magnolia fossil leaf.*
C. *Porifera, a sponge turned into flint and clay.*

This tiny bone, only just over 3 cm (1 inch) long and embedded in rock is the kind of evidence on which knowledge of the earliest mammals is based. It belonged to an Amblotherium, a small creature only a little larger than a present-day mouse.

LIFE IN A SPECK OF JELLY

No one can say for certain how, when or where life began on Earth. We do know, however, that the warmth and light of the Sun was essential for life, and therefore life could have begun only after the thick blanket of cloud that once permanently shut off the Sun from the Earth had been swept away.

When the Sun first shone on Earth, there was no soil and no plants. Indeed, there was not even any atmosphere as we know it today, so there could have been no life on land at that time.

All the signs point, therefore, to life having begun in the sea. What exactly caused it to happen we simply don't know, but many scientists now believe that it began as just a tiny speck of jelly called protoplasm.

It must have taken hundreds of millions of years for these specks of self-reproducing protoplasm to develop into complete cells. But when that happened, life made a giant step forward. For every living thing on Earth today, from a small garden weed to a giant tree, from an ant to an elephant, is made up of cells.

We cannot say exactly what these creatures were like, because they left no fossil record for us to examine. But we can assume that they were microscopically small and had thin, hair-like threads sticking out of their bodies which enabled them to swim. Such single cell creatures still exist today, and we call them flagellates.

In due time single cell creatures joined together to form a jelly-like mass. Still later this mass of cells merged more closely to form a living multi-cellular thing.

One thousand million years ago a simple living thing existed on the seabed. It was a sponge, and it was not very different from the sponges you can still see in the sea today.

Next came the medusa and the polyp. They had simple mouths and stomachs. From them, water worms and annelida, the ancestors of our insects, developed. The first living creature about which we really have complete knowledge appeared in the water about 400 million years ago, during the period known as Devonian. This was the Trilobite, a lobster-like creature which appeared in a variety of different forms. They had hard shells and many legs, but some had many eyes and some had none. Some appeared even to have no heads, while those that did had mouths on the underside.

Another strange sea creature of this period was the Nautiloid, the world's first true predator, or hunter. It had long tentacles to seize its prey as it crawled along the seabed in search of food.

As sea creatures grew bigger and bigger, one giant emerged, called the Stylonurus. It was three metres (ten feet) long and used its legs as paddles. It caught its prey with two pincer-like claws on its head and then stabbed it to death with its spiked tail.

THE FIRST CREATURE TO MOVE ON EARTH

Three hundred million years ago life in the sea was probably a motley mixture of armour-plated defensive fish and razor-teethed leviathans that preyed on everything within sight. It is interesting to note that the shark is today's best-known link with that savage past.

The largest predators of that period belonged to the group of arthrodires or joint-necked fish. Imagine a huge fish about ten metres (33 feet) long. A kind of hinge-joint connects its head to the fore-part of its body. With angry glaring eyes and great teeth it presents a formidable appearance as it streaks along in search of its prey.

The name of this voracious fish is Dinichthys and on the following two pages it is illustrated as it is about to seize a primitive shark, by name Cladoselache. This fish was about one metre (three feet) long so you can see what a huge fish the Dinichthys was.

While all this was happening, fish of a different kind were developing in the fresh water rivers and the great muddy pools that were etched in the still sparse foliage of Earth. One group of these fish were the Crossopterygii or lobe-finned fishes and they included the forebears of all the land-living back-boned animals – including man. One group is called the Rhipidistia and unlike the second group, the Coelacanthini, is today believed to be totally extinct.

It was the Rhipidistia that first emerged from the water to continue the chain of life on land in what we call the Devonian period, 300 million years ago. It is perhaps hard to imagine that a fish living in water would want to crawl out and start a new life on land. But it is unlikely when this happened that fishes wanted it.

The Earth at that time was a mass of swamps and rivers punctuating thick, dark forests. For a long time no rain fell, and the swamps and rivers where the primitive fresh water fish lived began to dry up.

Almost all these water creatures must have died. But a few survived. One fish that survived the prehistoric drought was the Eusthenopteron.

Eusthenopteron crawled from its muddy pool, took a deep breath and took oxygen from the air into its primitive lung. It dragged itself along for a few metres using its fins as legs. Slowly, very slowly, it was to

This splendid fossilized skeleton of a flying dinosaur known as the Pterodactylus was an exciting find for an eighteenth century scientist. Luckily for us, scientists are today searching all over the world for fresh fossils and new theories are always being formed about the appearance and habits of prehistoric animals.

The giant fish known as Dinichthys was ruler of the seas during the Devonian period. A fearsome predator, it is shown here just about to devour a small shark called the Cladoselache. Fossilized remains of the Dinichthys indicate that it was over 9 metres (30 feet) in length.

evolve into a successful land animal.

Eusthenopteron – a member of the Rhipidistia group and probably the first living creature to move on land – had another distinct advantage over the other fishes that were fighting the battle to survive. It was a carnivore, or flesh eater. The bodies of its dead and exhausted water companions ensured an abundant supply of food as it struggled to adapt itself to its new life and its new surroundings.

One of Eusthenopteron's relatives was the Coelacanth which was said to be extinct until, in 1938, one was caught near Madagascar by fishermen. Incredible as it seemed, here was a fish which belonged to a group which was one of the first to appear on Earth perhaps 3000 million years ago and which was thought to have been extinct for 50 million years.

Scientists all over the world were excited by the discovery, but, unhappily, study of the specimen was not possible, because it deteriorated quickly in the tropical heat.

Eusthenopteron was a fish that was to evolve into a land animal as centuries passed. A member of the Rhipidistia group it had a distinct advantage over other fish that were struggling to survive because it was a flesh-eater.

For years another live specimen was sought; and then, again off Madagascar, one was captured – a species presumed to be different from the first and now named *Malanie anjouanae*. Its colour was brilliant steel-blue, it weighed 20.36 kilos (100 lbs), was 3.87 metres (5 feet) long, and had three sets of fins which looked like stunted arms and legs.

Scientists were delighted, because their deductions made from the fossils of the Coelacanth were proved substantially correct by this living representative of a far-distant past, which had been drawn from the sea by sheer chance.

Was there, the palaeontologists wondered, the possibility that other pre-historic creatures were still alive in the depths of the sea?

Their hopes were justified when in 1938, the same year as the discovery of the first Coelacanth, the remains of another prehistoric land and water creature were found off the coast of Greenland. This was the Ichthyostega, a descendant of Eusthenopteron. Ichthyostega was an amphibian which provided the link between the Eusthenopteron, born to live in the water and emerged on to land, and the creatures that followed which were born to live on land *and* water.

About thirty million years after Eusthenopteron crawled out of his dried-up muddy swamp for the first time, all sorts of amphibians were roaming the dense forests on Earth.

This was the Carboniferous period (about 280 million years ago).

Let us suppose for a moment that you could step on to a magic carpet that will whisk you back in time, back through the hundreds of millions of years to this period. As you look about you, it will seem as if you were on another planet.

The first thing that would probably strike you would be the uncanny silence. Not only are the everyday sounds of the modern world – the rumble of motor traffic, factory hooters and the roar of aircraft overhead – completely absent, but you would not hear the twitter or chirp of a single bird, for in this world there are no birds. No dogs bark, no cats mew. There is not even the sound of a human voice.

A shallow sea ripples before you and on all sides and behind you stretch vast tracts of swampland.

The air is humid and still. The ground is covered with small green plants growing in the shade of high green tree-like plants, none of them resembling the trees of our modern world. The trees are immensely high, up to 30 metres (100 feet) and what is more they do not shed their leaves once a year. All the plants and trees about you keep their leaves all the year round, single leaves dropping from time to time.

No flowers bloom for the plants do not

have blossoms to produce their seeds.

To think, though, that apart from the plant life, no other form of life is living in this world would be a mistake, for in fact there are innumerable millions of insects everywhere, mayflies, centipedes, cockroaches and scorpions. Spanning the leaves of a nearby bush is a web made by a spider. Two giant dragonflies are shimmering across a distant pool of still water. Their wings are one metre (three feet) across.

Suddenly amongst the undergrowth a large body moves. At first you might think it is some type of alligator. It has a massive ugly head with bared cruel teeth. It is two metres (six feet) long and its movements, as it waddles towards the shallow sea, are slow and cumbrous. When it takes to the water, though, it glides easily along, propelled by its powerful tail and limbs. The name given to this large beast is Eryops. A Diplocaulus, usually to be found on the bottom of pools, dodges swiftly out of the way of the Eryops. The Diplocaulus has a strange wedge-shaped head (see page 13). It is one animal that survived into the next period, the Permian.

There was masses of food in the warm moist climate for all these insects and creatures.

Over many years it gradually became more attractive for amphibians, such as the Eryops, to spend more of their time on this green and pleasant land, where there was plenty of fresh vegetation to eat.

Nevertheless, the amphibian needed water. It had to keep its skin moist, just as an amphibian does today, and because its off-spring still lived as fish at the beginning of their lives, absorbing oxygen from water through their gills, it had to lay its eggs in water.

Now it becomes easier to understand the next stage in evolution.

Suppose an amphibian in the Carboniferous period laid its soft eggs in a muddy pool, and that before the eggs hatched it began to dry out. The eggs were not quite so covered, not quite so wet as they should be, but nonetheless they survived and hatched.

If the same process went on for a few million years, pools continually drying out before eggs were hatched, the eggs would develop shells to protect them on the out-

A relation of the Eusthenopteron (see page 28) was the Coelacanth which was, for a long time, believed to be extinct. There was sensation all over the world in 1938 when a live specimen of this fish was caught by fishermen near Madagascar.

side and membranes to shelter their unborn occupant on the inside. This system of protection evolved to allow the unborn creature to complete its development within its egg rather than in the water.

When the creatures hatched out of their shell-covered and membrane-endowed eggs, therefore, they did not need to begin life in water. Their land legs were ready to serve in the chase for food. It was born as a land creature.

It was, in fact, no longer an amphibian like the Eryops, but a reptile. This was a tremendous advance in the fascinating sequence of evolution.

Although in the later – (upper) – Carboniferous period, the first true reptiles appeared, we have not, up to the present time, any direct evidence that their eggs were of the shelled type. It is to the early years of the next period, the Permian, in fact, that we can ascribe the earliest known reptile egg.

We do know that during the time between the Carboniferous and Permian periods, the weather was changing and so the reptiles had to grow hard skin so that they would not be prostrated by heat or suffer agonies in the bitterly cold weather. Gone were the tranquil humid days. The Northern Hemisphere was becoming very cold, so cold indeed that huge areas of the world were covered with ice. For some time, too, mighty volcanoes had been erupting, huge earthquakes had been shaking the world and new mountain ranges were being thrown up. The seas of the Carboniferous period were divided and enormous salt lakes were formed.

This immense change took millions of years. Many creatures disappeared forever but Eryops and the Diplocaulus survived into the next – Permian – period.

Another survivor was the Mastodonsaurus – a true reptile – and a fish-eating animal.

We have now reached the point where the dinosaurs first begin to make their appearance. The stage is set so let us move quickly into this amazing and mysterious world.

An artist's impression of the Carboniferous landscape. Sullen and brooding, its silence is broken only by the rustle of moving insects and the whirr of a dragonfly's wings. Huge trees, unlike any growing today, abound everywhere. In the far distance, volcanoes are erupting for, despite the seeming tranquillity of this scene, the end of the Carboniferous period was a turbulent one when many physical changes took place.

The Permian Period

Several amphibians which had thrived during the Carboniferous period still survived at the beginning of the Permian period, which lasted from 270 million to 225 million years ago. But now the reptiles were taking command of the land and it is fair to say that the Permian period saw the beginning of the age of reptiles.

Amphibians (the word means 'double life') are creatures that live on land but lay their eggs in water. Frogs and toads are present-day examples. They lay their jelly-like eggs in lakes and ponds, and there the eggs hatch into tadpoles.

Reptiles, however, lay their eggs in hard shells and when their young hatch, there is no necessity for them to continue life in the water until they have sprouted legs, as is the case with tadpoles. They can live their lives on land.

There are plenty of reptiles living in our world today. Crocodiles, alligators and turtles are aquatic reptiles. On the other hand, lizards and snakes live where it is

dry. The one thing they have in common is that they all lay hard-shelled eggs.

We have already mentioned the many different types of insects prevalent during the Carboniferous period. There were many more diverse types in the world during the Permian age. Dragonflies and cockroaches still abound and in addition there were stone-flies, scorpion flies, grasshoppers, lacewings, beetles and flies similar to house-flies.

If your magic carpet could take you to the world of the Permian period, your attention would be immediately attracted to the great reptiles, such as the Mastodonsaurus.

Its head looked as though it was too large for its squat body and short legs. It must have lumbered along clumsily on land where probably it did not spend much time. We know that this big reptile – it was about 3.5 metres (12 feet) in length – lived on a fish diet. It rarely ate anything else. It follows, therefore, that most of its life was spent in the water and that if it was slow-moving on land, in the water it must have moved swiftly in order to catch its prey.

The Mastodonsaurus would bask lazily in shallow pools for lengthy periods and it was this habit that was one of the reasons why it became extinct during the next – Triassic – period. The pools where it lived often dried out and so the beast perished.

The Mastodonsaurus should not be confused with the Mastodon. The latter was an early elephant, which lived in prehistoric terms during the Tertiary period which was only one million years ago.

Excellent skeletons of the Mastodonsaurus have been unearthed and this is why we can formulate such a good impression of its appearance when alive.

The very early reptile group is named the Cotylosauria (meaning stem reptiles). The earliest of these were small flesh-eating animals dating back to the Carboniferous period. During the Permian period plant-eating Cotylosaurs appeared. They were Tetrapoda (meaning they walked on all four feet) and they are easily recognizable by their odd leaf-shaped teeth which were so different from the sharp fangs of the carnivorous reptiles.

The largest of these herbivorous reptiles, the first four-footed plant-eater, was the

This brooding beast perched on a huge rock is a Mastodonsaurus. It was about 3.5 metres (12 feet) long and its head was more than one metre (three feet long). Those comparatively small legs had a great weight to carry and it is likely that on land it was a clumsy animal. It lived almost entirely on fish and therefore it was probably a fast swimmer, preferring to spend most of its time close to pools where it could seize its prey with a minimum amount of effort. Pools, however, often dried out and then the unlucky Mastodonsaurus would die. It lived during the Permian period.

Pareiasaurus and in appearance it looked rather like a hornless rhinoceros. Its legs jutted at right-angles to its body and because of this it needed tremendously strong muscles and huge bones to support its weighty body.

A group closely related to the Cotylosauria is the Pelycosauria. The Pelycosauria contains both carnivorous and herbivorous types and some of these animals had greatly elongated spines on their backs and one such was the Edaphosaurus, seen here.

There has been a lot of discussion as to the purpose served by this strange structure. The most popular theory is that it was a heat-radiating device, which would have been a tremendous boon to a cold-blooded reptile like the Edaphosaurus.

Right: Edaphosaurus belongs to the Permian period. It was a long-spined plant-eating reptile, all of 3.5 metres (11 feet) long. It was a member of the Pelycosaur group of reptiles, animals which carried curious sail-like structures on their backs. Scientists today believe that this odd 'sail' was a device for controlling the body temperature of the creature, working like a radiator in getting rid of excess heat or alternatively absorbing heat during the early part of the day, thus enabling these cold-blooded animals to warm up and get moving.

Below: the skeleton of the Pareiasaurus shows the massive limb bones and the ungainly right-angled flexing of the legs.

It was a big beast some 3.5 metres (11 feet) long and weighed 250 kilos (555 lbs). The Edaphosaurus would have taken three and a half hours to warm up to its daytime temperature without a sail. But, because of its sail, it took only approximately one hour.

In the coldness of early morning the animal could turn its sail towards the sun and rapidly warm up. If, during the day, the sun's heat became intense, the Edaphosaurus could face into the sun and only the thin edge of the sail would be turned towards it thus cooling the rest of its huge body. This is only a theory, but it is widely held. Another group, called Therapsida, (meaning mammal-like reptile) evolved from the group of Pelycosaurs to which the Edaphosaurus belonged.

Prominent in this group was the Moschops, a craggy awkward brute whose skeleton measures 2.5 metres (8 feet) long. It was plant-eating and would have been unlikely to move far from pool sides and water holes.

Despite the formidable and threatening appearance of these reptiles, it is reasonably safe to say that they were harmless enough to other creatures as long as they were left alone. After all, they were not interested in the flesh of other inhabitants of their territory. This illustration shows a mother moschops with two young.

Before leaving the Permian period, let us take a look at one flesh-hungry reptile, known by the name of Sauroctonus.

Remains of this savage animal have been discovered primarily in the basin of the River Volga, in Russia.

200 and 300 million years ago, a group of animals that today we call Pelycosaurs appeared on Earth. They were four-footed plant-eaters with strange sail-like structures on their backs. From the Pelycosaurs there developed a second group of animals called Therapsida, mammal-like reptiles. Prominent amongst them was the Moschops, seen here with two of its young. Awkward and slow-moving, it fed on the plants growing beside water-holes and at the edge of pools. It was 2.5 metres (8 feet) in length.

w francis phillipps.

First, note the two pairs of canine teeth, sure pointers to the animal's liking for flesh. Once sunk into another creature's hide, those teeth would never be removed before they had torn away a huge jawful of flesh. Woe betide the unwary animal that crossed the path of the Sauroctonus. Its other teeth were smaller but just as sharp and pointed.

The Sauroctonus had parietal eyes. This means that it could move them independently, just like a modern chameleon. You can be sure that Sauroctonus was constantly moving in search of a victim.

On the very last pages at the back of this book you will see an illustration of a Pleuracanthus, a fresh-water shark of this period. Note the curious spine at the back of its head. It usually measured about 0.75 metres (2.5 feet).

As we have seen, the Permian period was one of great physical unrest in the world. Many upheavals had taken place due to the tremendous earthquakes and the face of the Earth had been vastly altered.

Now we are about to enter a new period, known as the Triassic, a period of 50 million years which saw the appearance of many new and remarkable animals, amongst them the first mammals and the first dinosaur.

This amazing evolution took place during a period of comparative calm, although the world had not yet seen the last of tremendous upheavals.

The ferocious teeth of the Sauroctonus, an animal of the late Permian period, tell that it was a flesh-eater. Its upper and lower jaws each contained a pair of massive canine fangs, the upper pair being the larger. The lower jaw of the beast widened to form a kind of chin. It was approximately three metres (ten feet) long and although its long five-toed limbs resembled a mammal's limbs, it was not an ancestor of the mammal.

The Triassic Period

The dawn of the Triassic period, which lasted for 45 million years, saw a world in which there were two great continents. The one in the north we now call Laurasia, and Gondwanaland was in the south. The continents were separated by a long belt of ocean which we call Tethys that stretched from the site of today's Mediterranean Sea thousands of miles eastward to the East Indies.

In the northern continent of Laurasia was most of what we now call North America, a part of the Atlantic Ocean, large sections of Europe and certain regions in Russia.

Gondwanaland included today's South America, Antarctica, Australia, India and Africa, also parts of the Middle East and Italy.

The two great continents were probably joined at a point midway between the western bulge of Africa and North and South America, not then, of course, covered by water.

Scattered across these vast territories were salt lakes, formed by the gigantic upheavals that had taken place during the turbulent Permian period.

Huge deserts and shrub-covered mountains formed most of the landscape and it was in these unprepossessing lands that the ancestor of all the dinosaurs was born.

It is, of course, well-known that many dinosaurs were flesh-eaters but which, you might ask, was the first? It is believed the one illustrated on the opposite page – Ornithosuchus – was the first. Today it is recognized in scientific circles as being the ancestor of all the large carnivorous dinosaurs – the fierce Allosaurus, the vicious Tyrannosaurus Rex and others – that strode the Earth during the later Jurassic and Cretaceous periods. During the later part of the Triassic period, the Ornithosuchus was the most formidable predator alive. In palaeontology it belongs to a group known as Thecodonts.

In appearance it was up to about three metres (nine feet) in length and one metre (three feet) high. Its jaws were furnished with razor-sharp, stabbing teeth. Its snout was very narrow, rather like a beak but very deep, and its jaws were actuated by powerful muscles.

The back and sides of its body were covered with tough bony plates. On its neck, these plates grew into sharp spines thus acting as a protection to this vulnerable part of the animal's anatomy.

As you can see, its forelegs were very short compared with its hind legs. Those long hind legs enabled the Ornithosuchus to pursue its prey at great speeds. When resting, it balanced on its long tail.

As the centuries passed, the descendants of the Ornithosuchus grew larger and larger, while the bony scales gave way to a scaly skin similar to the skin of today's lizards. Nature was, as usual, making the adjustments necessary to an animal's requirements. The only danger that existed so far as the Ornithosuchus and its descendants were concerned was from other dinosaurs of their own family, but these would only have concerned themselves with eating younger and smaller creatures.

Now we come to an animal that has proved valuable to scientists, for its remains have been found in the Antarctic, India and South Africa and so we know almost certainly that the continents of Laurasia and Gondwanaland were linked together. This beast is Lystrosaurus.

During the Upper Triassic period, 200 million years ago, a new reptile appeared. This was the Ornithosuchus. It is today recognized as the ancestor of the mighty flesh-eating dinosaurs that inhabited the Earth in the Jurassic and Cretaceous periods. Its skeleton indicates that it walked on only two legs, balancing the weight of its body on its strong tail. A carnivore, it possessed a fearsome array of sharp, stabbing teeth, ideal for seizing its prey.

Before the dinosaurs evolved, mammal-like reptiles, or para-mammals, were the most important form of life on land for more than 70 million years. At the beginning of the Triassic period, the para-mammals still dominated life on this planet. The Lystrosaurus was a plant-eating para-mammal which lived partly on land and partly in the water, similar to the life pattern of a modern-day hippopotamus. Its remains have been located in India, South Africa and the Antarctic. This is part of the evidence that leads us to believe that the continents were once joined together.

You will probably notice from the picture of the Lystrosaurus (opposite) that the animal bears a certain resemblance to a present-day hippopotamus.

In fact, it was semi-aquatic and it lived, much as the hippopotamus does, partly on land, partly in the water.

In spite of its baleful appearance, it was not a flesh-eater and this is confirmed by its small teeth. It was a mammal-like reptile, otherwise referred to as a paramammal. At the commencement of the Triassic period, it was this animal and other paramammals that were the most prominent animals then living. They were all herbivorous and, unhappily for them, they were to provide food for the flesh-eaters such as the Ornithosuchus and the Coelophysis which was a carnivorous dinosaur about the same size as the Ornithosuchus, but much more agile and graceful with a longer neck and tail.

Let us now look at the marine life of the time. To do this we will hark back to the Devonian period which preceded the Carboniferous period. It was then that many groups of bony fishes moved into fresh water. Many of these fishes, which had survived for nearly 200 million years, left the rivers and returned to the warm salty waters of the ocean of Tethys during the Triassic period.

What naturally followed was that the fish-eating reptiles pursued the fish into the ocean. They were compelled to adjust themselves to a new way of life and so we now come to a new group of reptiles, known as

During the Triassic period which followed the Permian, many new types of aquatic animals appeared, among them the Nothosaur. It was a rather long-necked and long-tailed reptile, varying in length from 30 cm (1 foot) to 6 metres (20 feet) and it could definitely move on land although its limbs were only moderately modified for swimming. Its digits were webbed. Note the rows of long sharp teeth. No fish grabbed by them could possibly have escaped.

lizard-fish. The scientific name for them is Ichthyosaurs. Most of these Ichthyosaurs measured up to three metres (ten feet) but some fossil remains discovered in Utah, USA, measured 13 metres (40 feet) in length. Other groups of marine reptiles were the Nothosaurs and the Placodonts.

Some reptiles during the Triassic period began to fly but it should be pointed out that the marine and flying reptiles were not dinosaurs.

The most prolific of all the marine reptiles were the Nothosaurs and to give some idea of how widespread they were, their remains have been found in England, France, Holland, Germany, Poland and Switzerland, China, Japan, India and Jordan. These countries were then covered by the Tethys Ocean. The illustration above

depicts one of these beautiful but deadly animals.

The Nothosaurs passed most of their lives in the sea, searching for the fish on which they lived. We know that they could live on land because the fossils of many young Nothosaurs have been found in caves and on land which was not then covered by the sea.

They had very long heads armed with lots of long sharp teeth. Once those cruel fangs were buried in the flesh of a fish, the unlucky victim would never have been able to escape. They had powerful webbed feet and as you can see their bodies were specially designed by Nature for gliding swiftly through the water.

They varied in length from 30 cm (12 in.) to 6 metres (29 feet).

The Nothosaurs were not Ichthyosaurs. They were a quite different species as were the Placodonts which looked rather like long-tailed tortoises. Placodonts lived on a diet of shell-fish and died out at the end of the Triassic period.

To return to the Ichthyosaurs, prominent in this group was the Mixosaurus. It is thought that this animal might have descended from the reptiles of the land but by the time of the Triassic period it could live only in the sea. It had a tail fin like a fish, its feet were webbed and its ugly teeth projected oddly from its mouth. The Mixosaurus was to survive and flourish in the next period which was to see the commencement of the Golden Age of the giant Dinosaurs. This was the Jurassic period, which was over 200 million years ago.

The Jurassic Period

The Jurassic period lasted 45 million years, from approximately 225 million years to 180 million years ago. Once again, great physical changes took place in the world. Huge land masses were divided up, and the numbers of shallow lakes and seas increased. It is difficult for us to conceive of periods of millions of years; but perhaps some idea can be grasped by the fact that what had been high mountains during the previous Triassic period had been worn down by the elements and in the Jurassic period were little more than low hills.

The weather during the Jurassic period was mild with lots of rain. So many plants and trees proliferated and new rivers and deltas were formed. The world was beginning to look something like our own world. Swampy plains were common and it was in this world that the first mighty dinosaurs thrived.

It would be wrong to assume that the word 'dinosaurs' covers all the animals that lived at this time. Dinosaurs were actually members of the class Reptilia and there are innumerable types of reptilia (both living and extinct) known to science.

The class Reptilia is subdivided into various groups of closely related classes. Snakes and lizards, for example, are brought together in the sub-class Lepidosauria. Turtles and some extinct reptiles are grouped in the sub-class Anapsida. Dinosaurs belong to yet another sub-class named Archosauria which means 'ruling reptiles'.

Archosauria are sub-divided yet again into five different orders – the Saurischia, the Ornithischia, the Crocodilia, the Pterosauria and the Thecodontia. (Ornithosuchus was a Thecodont.)

These five groups are sub-divided into many others, but for the purposes of this book we will go no further than the five groups mentioned above.

(It is interesting to note that present-day crocodiles more closely resemble dinosaurs than any other living reptiles, and birds have evolved from the same ancestors as dinosaurs. The Archaeopteryx was the first feathered dinosaur.)

Returning to the five different dinosaur sub-classes and to the Saurischia in particular, the word Saurischia means 'lizard-hipped', a term which is self explanatory. The Brachiosaurus, a sauropod, is illustrated on the opposite page. It was the biggest animal so far known about that ever walked on Earth.

In its lifetime it wallowed in shallow waters and swamps, much as you see it here, with its head just above the water. The skull around the nostrils forms a small dome. This was peculiar to this dinosaur.

Brachiosaurus means 'great arms' and it is so-named because its front legs (or 'arms') were longer and heavier than its back legs. It had a long neck surmounting a gigantic body, so huge that it needed four legs as big as Greek columns to support it. The animal must have moved along ponderously and slowly on land and would have preferred the easier movement it would enjoy when partially submerged in water. This was not the only reason why the Brachiosaurus had a predilection for the aquatic life. It was a herbivore and in spite of its colossal size – all of 23 metres (75 feet) with a head-height of 12 metres (40 feet) – left to itself it would have represented no danger to other animals. But carnivorous monsters were probably never very far away and the Brachiosaurus would fall a comparatively easy victim to one of these. Deep in water, with only its eyes peeping above, it would have felt reasonably safe from the onslaught of a carnivorous enemy.

Despite Brachiosaurus' huge size, it had a very tiny brain and it would not have been geared to any swift reaction to any form of imminent danger. Instinct probably prompted it to seek safety in hiding in water. There, the tasty weeds on the bed of the lake or pool could easily be reached by its long neck. Moreover, the water would keep it cool. What could have been better for such a mighty dim-witted monster?

If the Brachiosaurus was the biggest

Bigger still than the great Apatosaurus was the Brachiosaurus, seen here immersed in water and contentedly enjoying a meal. It was the biggest animal ever to appear on Earth. In Africa a skeleton of a Brachiosaurus was discovered, almost 23 metres (75 feet) long. Its shoulder height must have been in the region of 6 metres (20 feet) and therefore it would have stood approximately 12 metres (40 feet). Isolated bones of the Brachiosaurus have been found, though, indicating that the largest Brachiosaurs were considerably longer than the one found in Africa.

There is only one word to describe the size of the Diplodocus – colossal! The animal when alive must have stalked the Earth more than 26.5 metres (87 feet) long. A plant-eater, it represented no danger to other animals, although, of course, it was suitable prey for the great carnivores. It could fight hardily, though, for its tail formed an enormous whiplash, without doubt a formidable defensive weapon. The Diplodocus belonged to the group of dinosaurs known as Sauropeda. They all looked rather alike, short bodies, tall legs, long necks and tails.

beast of them all, the Diplodocus (meaning 'double-beamed') was the longest of all the dinosaurs at 28 metres (93 feet) long. Like its near relative, the Brachiosaurus, it had a small head with a brain no larger than a kitten's and a long neck. Its tail was 14 metres (46 feet) long. It was a herbivore, and almost defenceless despite its huge size. Its big tail was its only weapon and when its tiny brain managed to reason that danger was near it would lash out with it. Like the Brachiosaurus which also used its tail as a weapon, the Diplodocus, trapped on

land by a hungry carnivorous monster, would probably have succumbed without much of a struggle.

Unlike other vertebrates (animals with spinal columns) the Diplodocus and the Brachiosaurus, too, had legs of solid bone. Each of its back feet was equipped with five toes. The first three were horny-clawed and the other two toes were embedded in a powerful footpad, similar to the pad on an elephant's foot. Each front foot had only a single claw on the thumb, enabling the great animal to grip on muddy ground.

Overleaf is a magnificent illustration of a full-grown Apatosaurus. During the Jurassic period the ground would have trembled as one of these mighty animals strode along, for believe it or not, it weighed as much as five adult elephants. The Apatosaurus was one of the bigger Sauropods, walking on strong legs with short, broad feet. Many people call it the Brontosaurus but that is not its correct name. Apatosaurus means 'unreal lizard', probably because the scientist who found the first skeleton could not believe that it was real.

w francis phillipps.

The Jurassic period saw the first appearance of the mighty Stegosaurs, the armoured dinosaurs. The Scelidosaurus, seen here, was one of the earliest. It was about 3.5 metres (12 feet) long. We know that it possessed bony plates down its back but the precise arrangement is not known. With a small head, the weak jaw and grinding teeth of the plant-eating animals, it still managed to present a formidable appearance to the carnivores who preyed on it. From the Scelidosaurus evolved the great Stegosaurus (see pages 54 and 55) which were 6 metres (about 20 feet) long.

What did the Diplodocus eat? Probably nothing very substantial for its teeth were few and feeble. Its food therefore must have been very soft, possibly some kind of water plant.

Another huge animal of the Jurassic period was the one commonly referred to as the Brontosaurus, but as you have read already on page 6, scientists today call it the Apatosaurus. (It is illustrated on pages 50–51.) A sauropod, it reached a length of between 18 and 21 metres (60 and 70 feet) and weighed as much as five present day full-grown elephants.

Its powerful legs ended in short broad feet. There were three large claws on each of its hind feet, whereas each forefoot car-

ried only one. It had a large snaky neck, high enough for the beast to look over most trees. Like the Diplodocus, its teeth were small and good only for nibbling the softest plants. It may well have lived for as long as 200 years and it would have spent most of this time standing or ploughing through shallow lakes and rivers in search of food. All in all it was a harmless creature, despite its vast proportions, and it would not have been a very strong fighter when attacked by some hungry carnivore.

Buried in the top of its head was an air hole with a channel leading to its nostrils. This enabled the Apatosaurus to submerge in water and still breathe.

The animal above is a Scelidosaurus. It belonged to the second of the five sub-orders of Archosauria mentioned on page 47, namely the Ornithischia (bird-hip). The Stegosaurus and the Scelidosaurus belonged to this latter sub-order. It was a herbivore and measured about 3.7 metres (12 feet).

From the Scelidosaurus evolved the better-known Stegosaurus. On pages 54–5 you will see a breath-taking picture of a Stegosaurus unaware of the approach of a mighty Allosaurus bent on launching itself unexpectedly on the back of its spiky-backed victim.

But although the Stegosaurus was a plant-eater and the Allosaurus was a flesh-eater, the former might very well have

On the preceding pages is a graphic illustration of a Stegosaurus oblivious of the impending danger behind it – a hungry Allosaurus.

Below we see the last moments of the fight as the carnivorous Allosaurus leaps on to the Stegosaurus to kill it with its sharp pointed teeth. The Stegosaurus thrived mainly during the Jurassic period. The oddest thing about this dinosaur, which measured up to 6 metres (about 20 feet) in length, was its armament of two rows of large triangular bony plates along its neck and back and part of its tail. It was not armoured in any other way and was therefore open to an attack on its flank. How could it defend itself against the onslaught of some carnivorous enemy? Possibly by curling up like a hedgehog, head turned away from the would-be killer, while its plates were spread out and its tail lashed.

astonished its would-be killer by the cunning of its defence.

If its ancestor the Scelidosaurus was only 3.7 metres (12 feet) long, the Stegosaurus was much bigger in length, approximately six metres (20 feet) in length.

As you can see it had a very small head and inside that head was a brain no larger than a walnut.

One of its greatest aims in life for herbivores must have been to avoid the destructive and fearsome teeth of the various carnivores. This, of course, applied to all the herbivores but whereas the sauropods would take to the water in an endeavour to escape the flesh-eaters, other plant-eaters developed various forms of armour. All the dinosaurs' ancestors had bony plates in their skins but although many lost the plates, the Stegosaur's armour plating grew to large proportions, in two rows of triangular plates along the neck and back, and part of the tail. Approaching the end of the tail the plates were followed by two pairs of big spikes.

Stegosaurus, meaning 'roofed lizard', gets its name from its bony plates. Professor O C Marsh, who found the first fossils of the animal, thought that the plates looked like roof tiles.

It did not possess any other armour so it might seem to have been virtually defenceless to that attack of the Allosaurus. It did not have strong teeth and indeed the front of its jaw was toothless. (Like the sauropods, it probably lived on soft food and grasses.)

The fact of the Stegosaurus having such a small brain, however, has given it a notoriety that in fact it does not truly deserve.

While, more than likely, it was a beast of limited intelligence, it would be quite wrong to describe it as 'brainless'. All large reptiles have comparatively small brains. We have only to consider how small a crocodile's or a python's brain is to realize this, and even these present-day reptiles have some measure of reasoning ability. It would be reasonable enough to believe, therefore, that when a Stegosaurus and a Allosaurus fought each other, the former may well have proved a hardy opponent.

Those plates from their upright positions would probably have spread out horizontally while its tail lashed back and forth in an effort to knock its attacker off its balance. It would doubtless have striven to keep its head and neck well away from the daunting fangs of the Allosaurus and it is believed that it might have curled up like a hedgehog to protect itself.

Sometimes for the sake of effect, artists illustrating popular novels and books will illustrate a mighty Tyrannosaurus Rex of the Cretaceous period and a Stegosaurus engaged in mortal combat. Such a battle could never have taken place for when the tyrant dinosaur lived, the Stegosaurs had been extinct for millions of years.

The Allosaurus was an indefatigable hunter and footprints on a slab of hardened mud have been discovered in Texas, USA, indicating that one of these dreadful killers pursued its prey across the watery waste of a lagoon. Its great strides were over two metres (six feet) long. The Allosaurus was the biggest flesh-eating dinosaur of its time and it slew some of the biggest of all animals for food. Although its hind legs were so long, it was probably not a fast runner but it was more than speedy enough to overtake a lumbering Stegosaur or the gigantic Apatosaur.

It was about 9 metres (30 feet) long and its tail took up half its length. It had a massive head and a short powerful neck. It was a truly awesome beast and whereas several dinosaurs must have looked dangerous but were probably reasonably harmless, the Allosaurus almost certainly lived up to its frightening appearance.

Returning to the Ornithischians, one of the first members of this group was the Hyspilophodon and you can see one perched on the limb of a tree on the next page. What a very odd-looking creature it was. Its long name alludes to the strange shape of its teeth which had odd folds in them. Unlike the Stegosaur and other types of its group, it had teeth in the front of its jaw.

It was not one of the huge dinosaurs, being only about 1.5 metres (5 feet) in length.

The Hypsilophodon must have been a great survivor for its appearance remained unchanged for about 100 million years, from the time when the first dinosaurs evolved to the time when they uncannily vanished from the face of the earth. The skin on the Hypsilophodon's back still had

wfrancis phillipps.

faint traces of bony plates just like those first dinosaurs.

When fossil remains of this creature were first found, scientists concluded that it must have been a tree-dweller. It looked as though its big toe was set at an angle to the other three toes but now experts have decided that although the little finger of the animal's five-fingered hand protruded at right angles from the wrist, all the four toes of each foot pointed forwards.

While the long and strong fingers and toes would indicate that the Hypsilophodon spent a lot of its time perched on tree branches and fed on fruit and leaves (it is thought to have been herbivorous), it was probably not a tree-living animal.

Another Ornithischian, about three times larger than the Hypsilophodon, was the Camptosaurus. It probably walked on all its four legs except when running or reaching up to eat tree leaves. It was from the Camptosaurus that the mighty Iguanodon (see page 72) evolved.

The Hypsilophodon belonged to the dinosaur order known as Ornithischia. Several skeletons have been found in Southern England and in the form shown here, from the later Jurassic and Cretaceous periods, there are teeth in the front of the jaws. Its teeth suggest that it was herbivorous and because it must have resembled the present-day kangaroo, some scientists suggest that it might have climbed trees where it foraged for fruit and leaves in the foliage. It grew to a length of about 1.5 metres (5 feet).

THE FLYING LIZARDS

Now we come to the flying lizards and the fore-runners of the birds we see flying in the sky every day.

Flying lizards first appeared on earth at the beginning of the Triassic period and lived almost to the end of the Cretaceous period. They were probably extinct before the last dinosaurs disappeared. In all, they existed for about one million years and form the order known as Pterosauria.

The first Pterosaurs belong to the sub-order called Rhamphorhynchoidea. These aerial wonders had a long tail, in some cases, as shown in the illustration opposite, ending in a small rudder.

The wings of the Rhamphorhynchus were formed by the arms as in today's birds and bats. If you look closely at the picture opposite, you will see the 'hands and fingers' of the animal protruding from the leading edge of its wings. The leading edges from the hands to the tips of the wings are actually the immensely lengthened fourth finger of the animal. Extraordinarily, these very long fourth fingers were longer than the animal's arms. The three inner fingers corresponding to our thumb, fore and middle fingers (there was no fifth finger) were clawed, probably enabling the creature to hook itself up when resting. They may also have been used in shuffling or crawling along the ground.

The membrane of its wings were carried by the arms and long fourth fingers, extended to the hind legs, but the lower parts of the legs and feet were free of the membrane. Its five long toes were armed with curved claws which would have assisted the animal in clinging to its perching place. It does not seem likely that it could walk on its two legs as a bird does, for instance.

The flying lizards first appeared at the beginning of the Triassic period. Most people have heard of the Pterodactylus although this name probably refers to one group of Pterosauria. Pterosaurs are of two main kinds, the long-tailed Rhamphorhynchoids and the short-tailed or true Pterodactylus. The Rhamphorhynchus was representative of the former. It is depicted here. This group was confined to the Jurassic period.

The Dimorphodon was one of the most primitive Pterosaurs. Its large head dwarfed its comparatively small body and that head was equipped with rows of hard biting teeth. It also had large, strong hands and feet and a long naked tail.

Its body was small in comparison with its long wings. In startling contrast, its head was very big and its long slim powerful jaws were equipped with rows of viciously sharp teeth, pointing obliquely forward. It is named after the shape of its head for Rhamphorhynchus means 'prow-beak'.

Although it was not able to fly such long distances as a present-day bird, the Rhamphorhynchus was probably skilful in the art of gliding and soaring, taking every advantage of the winds prevailing at various altitudes. Its brain was moderately large and it is likely that it could well cope with the various problems associated with the art of gliding and swooping through the air.

As with most other dinosaurs and their predecessors, we are not sure of the facts concerning the origin of the flying lizards. The Dimorphodon, seen here, was also one of the first Pterosaurs.

Several of the Dimorphodon's characteristics, such as its strong hands and feet and its long tail, owed their origins to previous species but it was definitely a typical Pterosaur. One theory favoured by scientists is that the flying lizards evolved from tree-living, jumping Thecodonts whose wing membranes evolved to assist them in leaping and gliding from one branch to another.

Here dangling from a tree branch is a Pterodactylus, which means 'winged fingers'. The Pterodactylus was a true flying lizard and no relation to present-day birds. The form shown here was in size between a sparrow and a squirrel of today.

Pterodactylus is the name given to a sub-order that appeared during the late Jurassic period. This sub-order was more advanced than Pterosauria and as can be seen in the illustration opposite, its tail was so small as to be scarcely discernible.

The Pterosaurs were much bigger than the Pterodactylus. The Rhamphorhynchus, for example, had a wing-span of one metre (three feet) whereas the Pterodactylus was no larger than a present-day squirrel, and some were about the size of sparrows.

Pterodactylus had fewer teeth than Pterosauria and they were quite small. Thus we see the move towards the toothless beaks of today's birds.

Some long-jawed forms related to the Pterodactylus have closely spaced bristle-like teeth rather resembling the straining formation of the bills in certain present-day ducks. So once again we see the gradual evolution from the reptiles of millions of years ago to the birds we know so well today.

The Pterodactylus was probably quite an active little creature, able to wheel and soar over the lakes and lagoons, suddenly darting down to scoop up small fish and snap up insects in mid-flight.

How it moved on the ground is not certain but because of the shape of the wings it is believed that it probably slept as bats do, hanging downwards, suspended by its feet.

Not one of the various animals of the prehistoric ages that have so far been described, was clad in feathers. The first was the Archaeopteryx which translated means 'ancient wing'. It was the first kind of bird that appeared anywhere in the world.

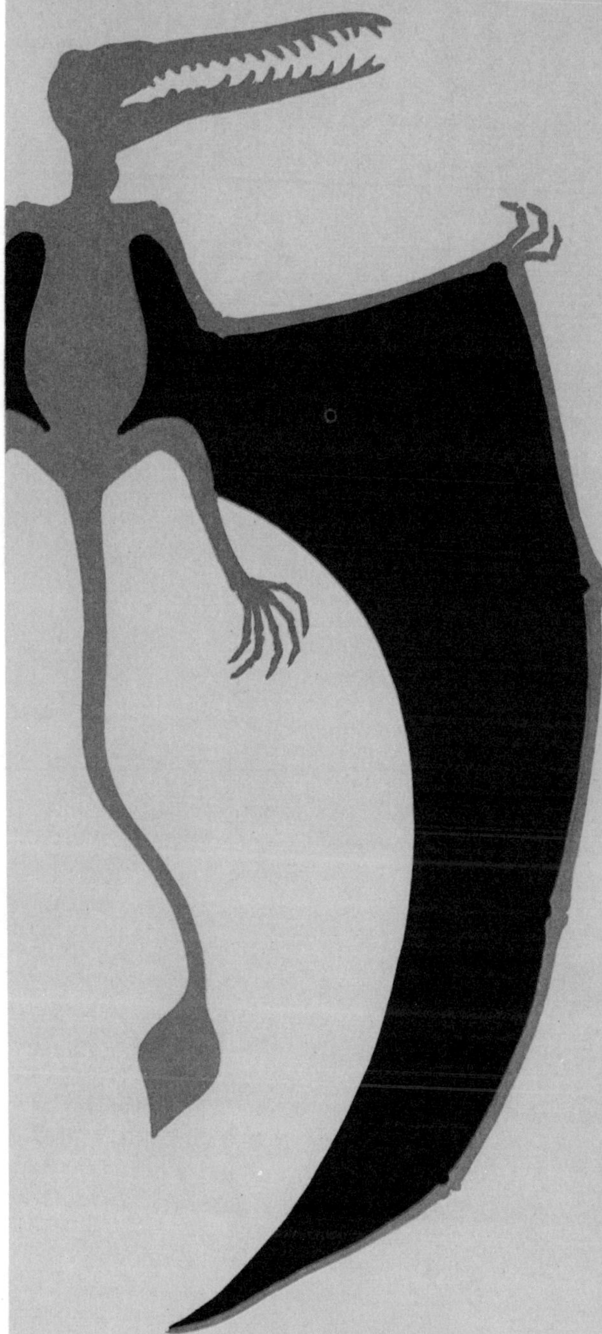

This diagram quite clearly shows the wing structure of a primitive flying lizard called Rhamphorynchus which translated means 'prow beak' (for the simple reason that its beak was shaped like the prow of a boat). The drawing clearly illustrates the incredibly long fourth finger of the animal extending from its hand to the tip of its wing.

It is believed that the Archaeopteryx was warm-blooded and therefore did not belong to the cold-blooded reptiles.

Even so, some reptilian features still remained. Its tail, for instance, was long and the feathers were set flat on either side, different from those of present-day birds. It still had teeth in its little reptile head, whereas today's birds do not. As can be seen in the accompanying illustration, it carried three spare claws on each of its wings and these enabled it to climb trees.

Although the archaeopteryx was plumed, it probably was not a good flyer. Its wings would have been far too weak to have lifted it from the ground. It is believed that to fly, it must have climbed a tree and then launched itself into mid-air and, taking advantage of the wind, soared and glided without much wing flapping.

It follows that when the Archaeopteryx alighted on the ground it was helpless to save itself from some predator, such as the Compsognathus, by flying away. Its only hope was to reach a tree fast enough to escape from a predator.

The Compsognathus is worthy of mention if only for the fact that it, so far as we know, was the smallest dinosaur of all. It was no bigger than a chicken, its head measuring 7.5 cm (3 inches). It had a long slim neck and its tail was longer than the whole of the rest of its anatomy. It is thought to have been carnivorous because in a skeleton once found in Bavaria, Germany, there was the skeleton of something that looked like a very tiny reptile. Flesh-eater though it probably was, its small frame would have made a suitable titbit for bigger carnivorous animals.

The Archaeopteryx represents a most important step in evolution. Present-day birds most likely evolved from tree-living reptiles that leaped from branch to branch. The Archaeopteryx was the first flying feathered *animal. It was about the size of a pigeon and although feathered, had reptilian scales on its legs and the typical long tail of a reptile. The small pointed teeth in its bill were of great significance.*

The Plesiosaur was a fast agile sea creature, highly adapted to life in the water. If it moved all its paddles forward, it was propelled backwards but by moving its paddles on one side in a different direction from the paddles on the other, it could turn swiftly, much as a rower can turn a boat. Plesiosaur means 'swan-lizard', and it was so named because of its long neck. Even so, certain types of Plesiosaur had short necks. They could not raise their 'flippers' above the level of their shoulder joints and so were unable to dive deeply after their food. For this reason they spend most of their time swimming on the surface of shallow waters.

MARINE LIFE OF THE JURASSIC PERIOD

Having dealt with some Jurassic animals of land and air, let us now turn our attention to marine life. The seas at that time were dominated by Ichthyosaurs, the fish-lizards.

The Ichthyosaurus looked very much like a fish and measured anything from one metre (three feet) to 12 metres (40 feet) in length. Across millions of years, it had evolved from a land-dwelling reptile which, during the Triassic period, had returned to the sea and could no longer move on land. Its legs had developed into fin-like paddles and it had a large triangular fin on its back. Its tail was also fin-shaped in appearance.

During the 19th century, several skeletons of the Ichthyosaurus were discovered in Germany. The skin of these skeletons was so well preserved in places that the pigment cells which give colour could be identified, and thus it is that we know that the backs of some of these creatures was a dark brown. Inside the skeletons were found the remains of a surface-water fish – Pholidophorus. Thus it is deduced that the Ichthyosaur was a fish eater that probably swam close to the surface of the sea.

We know that the Ichthyosaur did not lay eggs as a fossil was found which showed a mother and its young at the moment of

birth. Without doubt, the creature gave birth to its young in the sea.

The heads of many Ichthyosaurs were big with long, pointed jaws equipped with sharp teeth, although a few forms were toothless. (See illustration page 10.)

The sea beast pictured above is a Plesiosaur (meaning 'swan-lizard') so named because of its long snaky neck. It was descended from the Nothosaurs, the reptiles which had been so important on land and in the water during the previous – Triassic – period. Although designated 'swan-lizard', it looked more like a turtle with a long neck. However, not all Plesiosaurs were long-necked and the head, while looking quite small on the long neck, seemed quite large in the short-necked variety.

The long-necked Plesiosaurs were agile and fast and capable of intricate manoeuvres while the short-necked type were powerful swimmers that could swim for long distances.

Most animals move forward when swimming by thrusting their limbs backwards. The Plesiosaur was especially favoured inasmuch as the muscles responsible for the forward movement of its 'flippers' were almost as powerful as those actuating the backward movement. The animal could therefore move its paddles equally strongly in either direction.

Now, if you imagine a man sitting in a rowing boat who wants to turn his craft, you will see him push his oars in different directions. This action will effectively turn the boat. In this way could a Plesiosaur turn its body in a swift and instant turn.

The Plesiosaur was unable to raise its paddles higher than the level of its shoulder joints. This greatly restricted its movements so far as diving was concerned. Because of this, it could not dive down after its food and must have passed most of its life paddling about on the surface of shallow waters.

In the water, the long neck of the Plesiosaur moved from side to side like that of a snake but, while it could move its body rapidly to one side, it was unable to move its head at a comparable speed. As a result, when hunting, the Plesiosaur would keep its long neck out of the water and dart its head down to seize any prey that it sighted.

The Plesiosaurs that developed shorter necks during the beginning of the Jurassic period were the ancestors of a group called Pliosaurs. The Pliosaurs were not as manoeuvrable as the Plesiosaurs but could swim strongly for very long distances. They did not move their limbs, unlike the Plesiosaurs. Instead when moving through the water, they flapped their paddles much as a present-day penguin does. It might therefore be said that they 'flew' through the water.

Some of these creatures grew to lengths of over twelve metres (40 feet). It is possible that, strong-limbed as they were, they could move about on land, albeit somewhat clumsily. Possibly the Plesiosaur laid its eggs on land where they would have hatched.

Some Pliosaurs were whale-like in their immensity and it seems doubtful that such huge animals could have struggled out of the water to lay their eggs on land.

SEA CROCODILES

Today the only living animals which can with authority be traced back to the Archosaurs (or 'ruling reptiles') are the alligators and crocodiles. Indeed, in their fearsome appearance and accompanying ferocity they have changed very little in the millions of years since they evolved during the Triassic period.

One of the earliest forms of crocodile is named Teleosauridae. It had long slender jaws with vicious rows of teeth that bit into each other. Once those formidable teeth were sunk into the flesh of a victim, nothing could have saved it from being the next meal of the Teleosaur.

Pictured opposite is one of the savage members of the Teleosauridae. It is called Metriorhynchus. It is not difficult to believe that the present-day crocodile with its long snout and cruel teeth is a distant relative of this predatory sea beast.

With those strong limbs and mighty tail that could be lashed from side to side, it was a powerful and swift swimmer.

Wicked killer though it was, the Metriorhynchus was itself a prospective victim for other creatures of the sea – hunters that were seven or eight times the size of the Metriorhynchus, for example, the Liopleurodon, a Pliosaur whose head was three metres (ten feet) long, approximately the length of the Metriorhynchus.

With jaws agape, the Metriorhynchus streaks down through the water to seize an unwary victim. This dangerous killer was a member of a group of sea crocodiles called Teleosaurs. To help digest their food, they were in the habit of swallowing stones. Teleosaurs grew to the amazing length of fifteen metres (50 feet).

The Cretaceous Period

We now come to the Cretaceous period, in popular circles undoubtedly the most romantic of all prehistoric periods. The Cretaceous saw the emergence of the mighty monsters, Gorgosaurus, Tyrannosaurus Rex and the huge beast shown on these pages, the Iguanodon. To give you some idea of the vast size of the Iguanodon, a dinner party for 22 men was once held in a hollow life-size model of the animal and at the table the men, all of whom were comfortably seated, were all scientists who were attending an event which was being given in honour of the Iguanodon.

Why the Iguanodon particularly? Because it was the discovery of an Iguanodon's tooth in 1822 that started all the research and interest in dinosaurs. Until then nobody knew that such monstrous reptiles had ever lived.

This tree-high monster is the Iguanodon. It was the first dinosaur whose remains were discovered. Two or three species are known and seventeen skeletons of the largest were found in a coal-mine in Belgium. The animals when living were nearly 7.5 metres (25 feet) in length and about 3.5 metres (12 feet) in height when sitting up. So many Iguanodons meeting death at the same time would indicate that they used to move in herds. All the skeletons were adult. Young animals did not herd with their parents.

wfrancis phillipps

The Iguanodon was given its name because Dr Mantell, whose wife found the first Iguanodon tooth, thought that the fang looked like the tooth of an Iguana, a present-day South American animal.

Iguanodon belonged to a group of dinosaurs called Ornithopods, which means 'bird feet'. Tracks that have been found indicate that the Iguanodon could run quite fast. When it ran, it lifted its tail and sped along with its body inclining forward.

It was an herbivorous dinosaur and it probably used its great hands with their horny, spiky thumbs to pull down the branches of trees and eat the rich foliage as it is doing in our illustration.

Standing up the Iguanodon measured about 4.5 metres (15 feet) and from nose to tail more than 9 metres (30 feet) long.

If the plant-eating Iguanodon represented no great danger to other dinosaurs, the Gorgosaurus was a very different proposition for this gigantic terror, whose name means 'terrible reptile', was a carnivore, hungry for the flesh of any unlucky victim crossing its path. What a frightening figure it makes as it stands ferociously baring its teeth.

The Gorgosaurus and its close relatives the Tyrannosaurus Rex and the Tarbosaurus were the biggest flesh-eaters ever to live on land.

Here is a carnivorous animal that fully justifies its name of Gorgosaurus – terrible reptile. Gorgosaurus was almost as big as the Tyrannosaurus, which it preceded, and had long sharp teeth for snatching and tearing food, and bulky swift legs for running after its prey. Its forelegs were so short as to appear useless.

w francis phillipps.

The gigantic carnivorous dinosaurs such as the Tyrannosaurus Rex all belonged to the same species, the Carnosauria. They were prevalent during the Cretaceous period and by comparing this splendid illustration of a Tarbosaurus with those of the Gorgosaurus and Tyrannosaurus Rex on the preceding and following pages, you will note how similar they were in appearance, especially in regard to those two curious forefeet.

The Tarbosaurus opposite seems in all its power and arrogance to be brazenly defying the rain-storm drenching the barren landscape. Like the Gorgosaurus, it was a flesh-eater. Its name comes from a Russian word meaning 'terrible'.

The biggest of the Ornithopods was Tyrannosaurus Rex and the meaning of its name is ideally suitable – 'tyrant king'.

It was 16 metres (52 feet) long and stood 6 metres (20 feet) high. Its fore limbs, in comparison with its huge body, were tiny and useless for support, feeding or combat. But it more than made up for this lack with its great mouthful of razor-sharp teeth and powerful hind legs.

The illustration overleaf shows a mortal combat taking place between a Tyrannosaurus Rex and a Styracosaurus which was a horned plant-eating dinosaur. This animal had a huge horn on its nose and a bony shield with six long dangerous spikes backing its head.

Like Tyrannosaurus Rex, the Styracosaurus and other horned dinosaurs lived in the western territory of North America. A few fossils have been discovered in Asia but none in any other part of the world.

The last dinosaur sub-order to evolve was

Probably the most well-known dinosaur of them all is Tyrannosaurus Rex and this terrible scene shows this dread killer felling and tearing the life out of a Styracosaurus. It lived about 70 million years ago in the Western region of North America. It was like a creature from a horror film, 16 metres (52 feet) long and when standing up it was about 6 metres (20 feet) high. It was the biggest and most powerful flesh-eating beast that has ever lived and probably the most vicious killer as well. Tyrannosaurus Rex means 'king tyrant lizard', an apt description indeed.

Here is the king of the horned dinosaurs – the Triceratops. It is known that it grew to a length of about 6 metres (20 feet) and carried two big frontal horns (that curved outward and forward) as well as a nasal horn. Its neck shield was a solid, saddle-shaped plate with a jagged border. Its body was massive, its legs long and its tail powerful if somewhat short. Its cheek teeth consisted of a long series of teeth on top of each other with which the Triceratops did not chew in the ordinary sense. The upper and lower teeth slid vertically past each other like the blades of a pair of scissors. Such teeth are usually only found in carnivores. It is believed, however, that the Triceratops was a plant-eating dinosaur.

that Ceratopsia and the most primitive animal belonging to this species was the Protoceratops. This animal did not strut along on its hind legs but walked on all fours. (There is a picture of this animal on pages 20 and 21.)

In the later stages of the Cretaceous period an important group of Ceratopsians owed their ancestry to the Protoceratops. One member of this group was the animal known as Triceratops, meaning 'three horned face'. It is not difficult to see why it was endowed with this name.

Triceratops was a plant-eater and if left alone would have been a harmless enough creature. But those fearsome horns measuring one metre (three feet) were not there just for decoration. This dinosaur was a formidable fighter, a fact that has been proved by the scars on the fossilized skulls of other Triceratopsians that have been found from time to time. One such skull was discovered with one broken horn, surely the result of some desperate battle.

Compared with dinosaurs like Tyrannosaurus Rex, the Triceratops was not a big specimen. It was about 6 metres (20 feet) long and 3 metres (10 feet) high. To support the great weight of its body it had short sturdy legs and broad flat feet with blunt claws – three on the toes of its fore feet and four on the back.

An animal certainly not to be tampered with was the Triceratops. Although smaller than a Tyrannosaurus Rex, if attacked by one of those carnivorous monsters, it would have proved a dangerous opponent for the body of the Tyrannosaurus would have been vulnerable to those viciously spiked horns of the Triceratops. On all four legs the Triceratops would have charged and

buried its horns in the body of the Tyrannosaurus with such force that it is believable that it could have smashed the giant dinosaur to the ground and once that mighty frame had been hurled to the ground, the Tyrannosaurus would have been in mortal danger, being unable to rise.

It is conceivable, though, that such battles did not frequently take place for the Tyrannosaurus would have regarded the Triceratops with too much respect to attack it needlessly. Only a Tyrannosaurus which had not eaten for a long time and with no other prospect of food in sight would have

The illustrations on these pages show the Monoclonius, a horned dinosaur. Its name means 'one stemmed' and is an allusion to the single horn sprouting from its nose. It was about 5 metres (almost 16 feet) long and had a short neck shield. The other animals are Scolosauri, which were armoured dinosaurs belonging to the Ankylosaurus family. Its bony carapace with its great bony spikes and its spiked tail made it very difficult to attack, although it had a vulnerable soft under-belly.

dared to challenge a Triceratops.

Other horned dinosaurs were the Styracosaurus (about which animal we have already written) and the Monoclonius, shown below.

The Monoclonius had only one horn. It ante-dated the Triceratops by several million years. It was about 5 metres (16 feet) long and its horn, reminiscent of the horn on the nose of the present-day rhinoceros, was curved forward to make a weapon admirably designed for impaling an enemy.

Its name means 'one-stemmed' and refers, of course, to its single horn. Like its descendant, the Triceratops, it was herbivorous and a dangerous beast only when attacked by a carnivore.

It was during the Cretaceous period that a new species of dinosaur appeared. They were heavily armoured and the name given to them is Ankylosauria, that is 'crooked lizard' and refers to the curved ribs of the group.

Ankylosauria were descendants of the Stegosaurus but the species had evolved before the Stegosaurus had completely disappeared. Thus the two groups existed together for a considerable period.

The two armoured dinosaurs pictured below are both members of Ankylosauria. They are Scolosauri and were herbivorous animals. The Scolosaurus had the typically small head and little weak teeth of its species.

Note the blunt spikes on the back of this dinosaur, also the menacing barbed tail. This was no easy prey for any greedy carnivore; armoured as its sides and back were, the stomach of the Scolosaurus was soft. Moving along on its short, stumpy legs

though, with its body close to the ground, it could only have been attacked from this angle if it were turned over, not an easy feat.

Later species of Ankylosauria such as the Palaeoscincus, Nodosaurus and Ankylosaurus carried a heavier mail. Take the Ankylosaurus as an example. Its back was covered in bony plates and its sides were armed with tough spikes running the full length of its body. The end of the creature's tail was armed with a large cudgel. Thus, although its teeth were minimal and therefore useless if attacked, it was well protected by its leathery coat covered with bony plates and trimmed with pointed spikes and that unusual tail could without doubt deliver a bone-breaking blow to any unwary attacker.

If it was attacked, it probably flopped down on its vulnerable stomach and in this position its would-be killer would be confronted by a mass of bony plates.

A Gorgosaurus attempting to kill the Ankylosaurus for food would probably wind up with a mouthful of broken teeth after trying to bite through the armour of the Ankylosaurus.

Moreover, if the predator sustained a bone-smashing blow from the tail of the Ankylosaurus – it was, in fact, a large circular mass of bone – and as a result was felled to the ground, that animal was as good as dead for, if incapacitated, it would not be able to walk and forage for food and would starve to death.

Ankylosaurus was about 4.5 metres (15 feet) and between 1–1.5 metres (4–5 feet) high and although it was not ranked amongst the biggest dinosaurs its tough, solid body was as impregnable as a rock.

As you can see in the illustration, spiky bones protruded from its head and these spikes helped to protect its eyes. Amazingly, its eyes were further protected by bony shields on its eyelids.

It is interesting to note that nearly all the fossils of the Ankylosaurus that have been discovered were upside down when found. This may indicate that although it would appear to us that it would not have been an easy task for a carnosaur to overturn the Ankylosaurus, the flesh-eaters had devised some comparatively easy way of turning their prey over on to their backs. After all, it is a well-known fact that a turtle turned over on to its back is quite helpless. In such a position, the Ankylosaurus would simply thrash about, weakly kicking its legs, uselessly endeavouring to strike its enemy with its clubbed tail.

It would appear that in the days of the dinosaurs the world was well-populated with Ankylosaurs. There were many different species and they existed for almost sixty million years so they were obviously hardily surviving animals.

One of the most prolific dinosaurs in the continent that today we call North America was the Anatosaurus (meaning 'goose lizard'.)

As you can see from the picture overleaf it had a long narrow head, the mouth of which was shaped very much like that of a present-day duck. Hence its common description of 'duck-billed' dinosaur.

Opposite are two weird-looking dinosaurs, with their terrifying spikes and odd club-like tails. They are known as Ankylosauria, a name that really means 'crooked lizard' and refers to the strongly curved ribs. The Ankylosaurus was typical of one sort of armoured dinosaur. It had stout bony plates all over its back while its small head was protected by a thick bony roof to its skull. It was a plant-eater and despite the weak defence of its small head and weak jaws it lived throughout the entire eighty million years of the Cretaceous period. The length of its body could be up to about 4.5 metres (15 feet) with a shoulder height of up to 1.5 metres (5 feet). Its flailing clubbed tail provided Ankylosaurus with a good weapon when attacked.

wFrancis phillipps.

Scientists have named this animal Anatosaurus – goose lizard. It is more often referred to as 'the duck-billed dinosaur'. On land it walked on its two sturdy back legs but often dropped to all four feet to probe along the ground and in the mud with its broad bill for roots and bulbs and snails. The Anatosaurus was most likely the best of the dinosaur swimmers. Its front feet were webbed and it swam by paddling with these feet and swinging its tail from side to side.

The Anatosaurus belonged to an important group of ornithopod dinosaurs named Hadrosauridae. (The word Hadrosaurus means 'big reptile.') Many skeletons of this animal have been discovered in North America, others in Europe and Asia.

It measured some 9 metres (30 feet) long and the formation of its body was rather like that of the Iguanodon.

Apart from the Anatosaurus, other Hadrosaurs were the Corythosaurus, the Lambeosaurus and the Parasaurolophus. These three latter Hadrosaurs had bony 'crests' on their heads. The Anatosaurus did not have such a crest but its other characteristics were similar to all Hadrosaurs.

They walked on two feet and each foot had three rounded hoofs. Their strong arms ended with three fingers with little hoofs. Some excellent mummified skeletons of these dinosaurs have been discovered so well preserved that we know what their insides and outsides looked like.

Those duck-bills were full of teeth. Each side of the upper and lower jaw had from two to three hundred teeth, so in total the Hadrosaur possessed about a thousand teeth. The stomachs of the mummified skeletons contained many different seeds

Overleaf is a picture of three dinosaurs resembling ostriches menaced by a hungry carnosaur. This ostrich-dinosaur is Struthiomimus and the name simply means 'ostrich imitator'.

wfrancis phillipps

A Pteranodon greedily snatches up a fish but as it seizes its prey it must keep a keen look-out for the huge carnivorous animals of the sea which will kill and eat a Pteranodon should it come within striking distance. Such a carnosaur might well be the rapacious Tylosaur (see pages 92 and 93). Swift in flight, the Pteranodon was awkward on earth for its hind legs were weak and quite short. It was probably a warm-blooded creature, in spite of the fact that it was a reptile.

and fruits, small twigs and pine needles, indicating that the animal was herbivorous.

The odd-looking creatures on page 88 in form resembled ostriches with long tails. That was why when deciding on a name scientists called the animal Struthiomimus – 'ostrich imitator'.

It was the same height as the present-day ostrich, about 2.4 metres (8 feet). The bill of the Struthiomimus was toothless, similar to that of an ostrich.

The long legs of the ostrich enable it to run fast. Doubtless the Struthiomimus with its long hind legs was also given to rapid turns of speed.

It is believed that the Struthiomimus may have eaten the eggs of other animals. Egg-laying animals abounded in the Cretaceous period and the Struthiomimus would have been able, with its hands, to snatch eggs from nests and then crack the shells with its jaws. Its other diet was made up of worms, insects and lizards.

It's predilection for eggs is thought to be confirmed by the fact that in Mongolia the crushed skull of a Struthiomimus was

found amid the broken eggs of another species. It was probably eating the eggs when it met its death, some way or other.

The scene below shows a Pteranodon whipping a fish up from the surface of the sea. Pteranodon, meaning 'winged, toothless', had a wing span of up to 8 metres (25 feet) and was the largest known flying reptile. Its horny beak was completely toothless and its large head looked even larger by reason of a long crest, which probably acted as a counterweight to the bill. The Pteranodon was the last flying lizard on earth. It provided food for carnivorous inhabitants of the sea such as the Tylosaurus illustrated overleaf.

The Tylosaurus was a mosasaur or sea monitor, closely related to the monitors of today, the largest of which, the Komodo monitor, measures three metres (ten feet). The Tylosaurus was much larger, up to 15 metres (50 feet). Scars on the fossils discovered of this sea beast tend to verify the theory that it was a vicious predator.

The illustration at the very front of this book show the Tylosaurus in action once again – this time it is attacking an Elasmo-

The picture on the following pages shows what happened when a Pteranodon has come too close to the surface of the sea in search of food and has been spotted by a hungry Tylosaurus. The Tylosaurus was a mosasaur or sea monitor with a large head, short neck and a long body and tail. When people talk about sea serpents they probably have something like the Tylosaurus in mind. It certainly was a most vicious carnosaur and desperate fighter as scars on its fossils clearly show.

wfrancis phillipps

saurus which grew to a length of about 14 metres (45 feet) and was very well represented in the sea during the Cretaceous period, following the disappearance of the Plesiosaurs.

THE UNSOLVED MYSTERY

For nearly 200 million years the dinosaurs were masters of the Earth. Then, about 70 million years ago, they disappeared. Now we only know that they existed from their fossil remains.

Just why they were completely swept away after roaming the Earth for 200 million years is a mystery. Many theories have been based on the fact that over the centuries the Earth has gone through many changes of climate.

Certainly the world that the dinosaurs knew was a vastly different place. All the land, even the Antarctic, sweltered under a humid, tropical heat. There were dense forests and vast plantations of giant ferns and these provided the enormous amounts of the green food on which most dinosaurs lived.

Then the weather gradually became colder and the rich, tropical vegetation began to disappear. According to some scientists, this meant that the dinosaurs starved to death and so became extinct. The only survivors were a few species small enough to live on the vegetation that remained. Today's lizards are descended from these survivors. But this theory had one flaw.

Despite the change in climate there were still tropical regions with rich vegetation where dinosaurs could have survived. Even today, there are tropical regions which could support a small population of dinosaurs.

Another theory to account for the dinosaurs' extinction is that the great creatures fell victims to a world-wide epidemic of some deadly disease. Whatever the epidemic may have been, it did not attack the animals' bones. If it had, there would be some sign of it in dinosaur fossils.

In any event, no animal species of modern times has ever been wiped out by disease. Species have been attacked by sudden and deadly epidemics and huge numbers have died. But the species has always developed a natural resistance to the disease and the losses were soon made up.

So even if some disease had killed many dinosaurs, it is unlikely that it could have wiped out the whole species.

Another theory is that when the world's temperature dropped, the dinosaurs died from the cold. Many zoologists believe that dinosaurs were cold-blooded. That is, they could not adjust their body temperature to changes in climate. In other words, the dinosaurs froze to death. But if one accepts this theory, it is still difficult to understand why some dinosaurs did not survive in the warmer parts of the world.

Yet another school of scientists think that the giant reptiles were warm-blooded. In that event, they would have been able to adjust to temperature changes and some at least would have survived.

Like most reptiles today, the dinosaurs laid eggs from which the young were hatched. This has caused some scientists to think that the long reign of the dinosaurs ended because their eggs were eaten by other animals.

The female dinosaur laid only a few eggs during her lifetime, and they were all the more vulnerable because she left them in the sun to hatch.

This theory is not so far-fetched when we remember what is now happening to the marine turtle. Eggs of the marine turtle are being collected in such vast quantities for human consumption that the creature is coming dangerously near to extinction.

According to the latest attempt to solve the mystery, the dinosaurs were certainly warm-blooded animals and might have adjusted themselves to survive in a colder world but for two factors: their enormous size and the fact that their bodies were covered with scales instead of fur.

Fur would have kept the creatures warm when the climate became cold, and the vast size of the animal presented such a huge surface to the cold that it just could not keep warm.

Even that theory has its flaws. Not all species of dinosaurs were gigantic. Yet the small species disappeared just as completely as did their gigantic relatives.

Whatever did happen to the dinosaurs, their complete disappearance from the Earth remains a puzzling mystery.

Glossary

Allosaurus, 6, 7, **8**, 40, 53, **54**, **56**, 57
(AL-lo-SAWR-us)
Amphibian, 29, 30, 32
(am-FIB-i-an)
Anapsida, 46
(an-APS-id-a)
Anatosaurus, 84, **86**, 87
(a-NAT-o-SAWR-us)
Ankylosaurus, 82, 83, 84, **85**
(ANG-kil-o-SAWR-us)
Apatosaurus, **6**, 7, **8**, 10, **50**, 52, 57
(a-PAT-o-SAWR-us)
Archaeopteryx, **11**, **25**, 46, **66**, 67
(ARK-i-OP-ter-ix)
Archosauria, 46, 53, 70
(ARK-o-SAWR-ia)
Arthrodires, 25
(AR-thro-dire-s)

Brachiosaurus, 46, **47**, 48
(BRAK-io-SAWR-us)
Brontosaurus, **6**, **50**, 52
(BRONT-o-SAWR-us)

Camptosaurus, 60
(KAMP-to-SAWR-us)
Carboniferous, 28, **30**, 32, 43
(KAR-bon-IF-er-us)
Ceratopsia, 80
(Ser-a-TOPS-ia)
Chart of eras and periods, **12**
Cladoselache, 25
(clad-o-SELL-a-key)
Coelacanth, 25, 28, **29**
(SEEL-a-kanth)
Coelophysis, 43
(seel-o-FIE-sis)
Compsognathus, 67
(COMP-sog-NATE-us)
Corythosaurus, 87
(ko-RITH-o-SAWR-us)
Cotylosauria, 33, 34
(ko-TILE-o-SAWR-ia)
Cretaceous, 40, 57, 60, 72, 82, 90, 94
(kree-TAY-shuss)

Devonian, 24, 43
(di-VONE-ian)
Dimorphodon, **62**, 65
(die-MORF-o-don)
Dinichthys, 25, **26**
(die-NICK-this)
Dinosaur, 14, 40, 94
(DINE-o-sawr)
Diplocaulus, **13**, 29, 30
(dip-lo-CAWL-us)
Diplodocus, **48**, 49, 52
(dip-lo-DOK-us)

Edaphosaurus, 34, **35**, 36
(e-DAF-o-SAWR-us)
Elasmosaurus, Front endpapers, 94
(e-LAZ-mo-SAWR-us)
Eryops, 29, 30
(ERR-ee-ops)
Eusthenopteron, 26, **28**
(yoos-theen-OP-ter-on)

Gorgosaurus, 72, **74**, 77, 84
(gorgo-SAWR-us)

Hadrosaurus, 15, 84, 87
(had-ro-SAWR-us)
Hypsilophodon, 57, **58**, 59
(HIP-si-LOFF-o-don)

Ichthyosaur, **10**, 44, 68, 69
(IK-thee-o-SAWR)
Ichthyostega, 28
(IK-thee-o-STEEG-a)
Iguanodon, **14**, 15, 60, 72, **73**, 74, 87
(ig-yoo-AHN-o-don)

Jurassic, 10, 40, 46, 60, 65, 68, 70
(joo-RASS-ik)

Komodo monitor, 91
(ko-MODE-o)

Lambeosaurus, 87
(LAM-be-o-SAWR-us)
Lepidosauria, 46
(le-PID-o-SAWR-ia)
Liopleurodon, 70
(lie-o-PLOOR-o-don)
Lystrosaurus, 40, **42**, 43
(LIS-tro-SAWR-us)

Mastodon, 33
(MAST-o-DON)